T0210901

SpringerBriefs in Mathematical Physics

Volume 17

More information about this series at http://www.springer.com/series/11953

Akihito Hora

The Limit Shape Problem for Ensembles of Young Diagrams

 Springer

Akihito Hora
Department of Mathematics
Hokkaido University
Sapporo, Hokkaido
Japan

ISSN 2197-1757   ISSN 2197-1765   (electronic)
SpringerBriefs in Mathematical Physics
ISBN 978-4-431-56485-0   ISBN 978-4-431-56487-4   (eBook)
DOI 10.1007/978-4-431-56487-4

Library of Congress Control Number: 2016955519

Printed on acid-free paper

This Springer imprint is published by Springer Nature
The registered company is Springer Japan KK
The registered company address is: Chiyoda First Bldg. East, 3-8-1 Nishi-Kanda, Chiyoda-ku, Tokyo
101-0065, Japan

Preface

Imagine a large statistical ensemble of Young diagrams and pick up one. We would like to say something about the typical shape, if any, of a Young diagram we get. Mathematically, let $\mathbb{Y}_n$ be the set of Young diagrams of size n and introduce a probability $\mathbb{M}^{(n)}$ on $\mathbb{Y}_n$. We discuss probabilistic limit theorems, especially the law of large numbers, as $n \to \infty$ on the quantities describing the shape of a Young diagram. While a Young diagram grows with n, let us rescale it horizontally and vertically by $1/\sqrt{n}$ to keep its area, which enables us to recognize the visible limit shape. Among others, the Plancherel measure is the most important from the point of view of symmetry or group-theoretical meaning. It describes the relative size of each irreducible component in the bi-regular representation of a symmetric group. Moreover, because the Plancherel measure is defined also on the path space of the Young graph, we can discuss the limit shape of Young diagrams as a strong law of large numbers.

Such a limit shape problem for Young diagrams was first shown and solved by Vershik–Kerov [29] and Logan–Shepp [21]. Afterwards, Biane [1, 2] extended this problem to a wide range of group-theoretical ensembles and brought in new insights of Voiculescu's free probability theory. Analysis of Young diagram ensembles and random permutations has made great progress, strongly influenced by an explosive development of random matrix theory. Beyond the law of large numbers, the central limit theorem (fluctuation of the shape) and other limit theorems have been studied extensively. References would be too huge to mention here (Kerov's book [19] is the one I always cite as a rich source of ideas from asymptotic representation theory). Readers can search through keywords and researchers according to their tastes.

This book is intended to serve as an introduction to the limit shape problem for Young diagrams as sketched above. It does not cover a broad range but stays near the classical results of Vershik–Kerov and Logan–Shepp. However, we bring a contemporary point of view for methods of proofs and some approaches. A key ingredient will be the algebra of polynomial functions in several coordinates of Young diagrams, which was introduced by Kerov–Olshanski [20]. In this book, we call it the Kerov–Olshanski algebra (KO algebra) after [20]. We give complete and self-contained proofs to the main results within the framework of representations of symmetric groups, not relying on random matrix theory or representations of unitary groups. Another point put anew is to mention a dynamical model for the time evolution of profiles of random Young diagrams. Although we focus mostly on the representation–theoretical aspect of the model in this book, analysis of the time evolution of profiles will be a promising topic with relation to geometric partial differential equations.

It is essential to investigate in detail the relations between various generating systems of the KO algebra, which was performed by Ivanov–Olshanski [16]. Notions of free probability theory are brought into this algebra with the help of Kerov's transition measure, and Biane's method plays an active part therein. Actually, it may be an exaggeration that we bring in the KO algebra to show the classical result of Vershik–Kerov and Logan–Shepp on the limit shape with respect to the Plancherel measure. However, once we know some structure of this algebra, the rest will be reduced to a pleasant application of simple weight counting argument. The KO algebra is a very nice device having rich applications in asymptotic representation theory for symmetric groups, especially in that it enables us to proceed along an exact or non-asymptotic way up to certain stages. We willingly include some materials about the KO algebra in reasonable depth. Such being the case, this book owes much to the works of [2, 3, 16].

Because the scope of this book is kept rather limited, we let quite many materials drop out of the content which could be appropriately included as interesting related topics by a more skillful author; for example,

- the philosophical and phenomenological analogy between random permutations and random matrices
- exact and asymptotic analysis of random Young diagrams as a point process
- the nature of fluctuations for ensembles of Young diagrams
- harmonic and stochastic analysis on infinite-dimensional dual objects, e.g., the Martin boundary of a branching graph
- asymptotic representation theory in frameworks beyond group actions, e.g., an extension from Plancherel to Jack, and so on.

Let us briefly give the organization of the following chapters. Because Chap. 1 is nothing but a casual description of preliminaries, readers should look into appropriate references according to their backgrounds. Speaking of representations of the symmetric group, one can go ahead with little trouble by accepting the hook formula and Frobenius's character formula. Chapter 2 is devoted to analysis of the KO algebra, which makes a technical prop. Chapter 3 contains analytic descriptions of continuous diagrams, or continuous limits of Young diagrams. Solutions of the limit shape problem for the Plancherel ensemble are given in Chap. 4. We give the proofs not only by an application of the KO algebra but also through what is called a continuous hook. The latter is of interest leading to the large deviation principle. While the results in Chap. 4 are of static nature, Chap. 5 includes a dynamical model. Funaki–Sasada [11] treated hydrodynamic limit for evolution of the profiles of Young diagrams. Chapter 5 is based on [12], which was greatly inspired by [11].

Sapporo, Japan   Akihito Hora

Contents

Chapter 1
Preliminaries

Abstract In this chapter, we briefly sketch the following materials as preliminaries for later chapters: representations of the symmetric group and Young diagrams, the Young graph and the Thoma simplex, combinatorial aspects of free probability theory.

1.1 Representations of Symmetric Groups

It is expected that our readers are either familiar with elementary terms of representations of (finite) groups and what we note in this section, or willing to take them for granted as well-known facts.

Young Diagrams

A Young diagram λ of size $n \in \mathbb{N}$ is specified by non-increasing integers: $\lambda_1 \geq \lambda_2 \geq \cdots \geq \lambda_{l(\lambda)} > 0$ such that $|\lambda| = \sum_{i=1}^{l(\lambda)} \lambda_i = n$, where λ_i is considered as the length of the ith row and $l(\lambda)$ is the number of rows of λ. Alternatively, λ is expressed as $(1^{m_1(\lambda)} 2^{m_2(\lambda)} \ldots j^{m_j(\lambda)} \ldots)$ by letting $m_j(\lambda)$ denote the number of rows of length j. The set of Young diagrams of size n is denoted by $\mathbb{Y}_n$. A Young diagram is displayed by loaded boxes or cells as in Fig. 1.1.[1] The box lying in the ith row and jth column is referred to as the (i, j) box. The transposed diagram of λ is denoted by λ'. The number of columns of λ then agrees with $l(\lambda')$.

Given $\lambda \in \mathbb{Y}_n$, a tableau of shape λ is an array of $\{1, 2, \ldots, n\}$ put into the n boxes of λ one by one. A tableau is said to be standard if the arrays are increasing along every row and column. The set of tableaux of shape λ is denoted by $\mathrm{Tab}(\lambda)$. As a subset we set $\mathrm{STab}(\lambda) = \{T \in \mathrm{Tab}(\lambda) | T \text{ is standard}\}$. The following formula counting $|\mathrm{STab}(\lambda)|$ is well-known. Here $h_\lambda(b) = \lambda_i - i + \lambda'_j - j + 1$ is the hook length of the (i, j) box in λ as it looks like in Fig. 1.2.

---

[1] In this book, we will have a Young diagram in the English style in mind for a combinatorial or counting argument. On the other hand, we will switch the picture to the style in Fig. 2.1 introduced later (often referred to as the Russian style) when some coordinates and profiles are treated.

© The Author(s) 2016
A. Hora, *The Limit Shape Problem for Ensembles of Young Diagrams*,
SpringerBriefs in Mathematical Physics, DOI 10.1007/978-4-431-56487-4_1

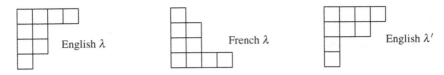

Fig. 1.1 $\lambda = (4, 2, 2, 1) = (1^1 2^2 3^0 4^1)$, $\lambda' = (4, 3, 1, 1)$

Fig. 1.2 (*left*) b: $(2, 1)$ box,
$h_\lambda(b) = 4$; (*right*) b: $(2, 2)$
box, $h_\lambda(b) = 2$

Proposition 1.1 (Hook formula) *The number of the standard tableaux of shape λ is given by*

$$|\mathrm{STab}(\lambda)| = n! \Big/ \prod_{b \in \lambda} h_\lambda(b), \qquad \lambda \in \mathbb{Y}_n.$$

Symmetric Groups

The symmetric group $\mathfrak{S}_n$ is the group consisting of the permutations of n letters $\{1, 2, \ldots, n\}$. We have an increasing family

$$\{e\} = \mathfrak{S}_1 \subset \mathfrak{S}_2 \subset \cdots \subset \cdots \subset \mathfrak{S}_n \subset \cdots \tag{1.1}$$

by regarding $\mathfrak{S}_m$ as the stabilizer of letters $m + 1, \ldots, n$ in $\mathfrak{S}_n$ for $m < n$. The unfixed letters for the action of $x \in \mathfrak{S}_n$ is called the support of x: $\operatorname{supp} x = \{i \in \{1, 2, \ldots, n\} | x(i) \neq i\}$. The support of x is well-defined along with the inclusion (1.1). Every $x \in \mathfrak{S}_n$ is decomposed into a product of disjoint (hence commutative) cycles, which assigns to x a cycle type $\rho = (\rho_1 \geqq \rho_2 \geqq \cdots) \in \mathbb{Y}_n$ where ρ_i's are the cycle lengths. Two $x, y \in \mathfrak{S}_n$ have the same cycle type if and only if x and y are conjugate. Let C_ρ denote the conjugacy class in $\mathfrak{S}_n$ consisting of the elements of cycle type $\rho \in \mathbb{Y}_n$. It is easy to see that

$$|C_\rho| = n!/z_\rho \quad \text{where} \quad z_\rho = \prod_j j^{m_j(\rho)} m_j(\rho)!. \tag{1.2}$$

Irreducible Representations of $\mathfrak{S}_n$

Several ways are well-known to assign an irreducible representation of $\mathfrak{S}_n$ to $\lambda \in \mathbb{Y}_n$ and to show that $\mathbb{Y}_n$ parametrizes the equivalence classes of irreducible representations of $\mathfrak{S}_n$. A recipe based on the action on the Specht polynomials is as follows. Set

$$\Delta(x_1, \ldots, x_n) = \prod_{1 \leqq i < j \leqq n} (x_i - x_j) = \det[x_i^{n-j}]_{i,j=1}^n.$$

If $\lambda \in \mathbb{Y}_n$ is a one-column diagram, then for $T \in \text{Tab}(\lambda)$ filled with letters $i_1, i_2, \ldots$ from the top we set $\Delta(T) = \Delta(x_{i_1}, x_{i_2}, \ldots)$. If $\lambda \in \mathbb{Y}_n$ is a general shape, then for $T \in \text{Tab}(\lambda)$ with T_j as the jth column we set $\Delta(T) = \Delta(T_1) \cdots \Delta(T_{l(\lambda')})$. The actions of $g \in \mathfrak{S}_n$ on tableau T and polynomial $F(x_1, \ldots, x_n)$ are defined by

$$(gT)(i, j) = T(g(i), g(j)), \qquad (gF)(x_1, \ldots, x_n) = F(x_{g(1)}, \ldots, x_{g(n)}). \quad (1.3)$$

Here $T(i, j)$ denotes the letter put in the (i, j) box in tableau T. Since $\Delta(gT) = g\Delta(T)$ holds, $\{\Delta(T) | T \in \text{Tab}(\lambda)\}$ spans an $\mathfrak{S}_n$-invariant subspace which is called a Specht module and denoted by S_λ. Restricting the action of (1.3) to S_λ, we get a representation (π_λ, S_λ) of $\mathfrak{S}_n$.

Proposition 1.2 *The set $\{\Delta(T) | T \in \text{STab}(\lambda)\}$ forms a basis of S_λ. In particular, $\dim S_\lambda = |\text{STab}(\lambda)|$.*

If $\mu \in \mathbb{Y}_{n-1}$ is obtained by removing one of the corners of $\lambda \in \mathbb{Y}_n$, we write as $\mu \nearrow \lambda$. We can show the decomposition

$$\text{Res}^{\mathfrak{S}_n}_{\mathfrak{S}_{n-1}} \pi_\lambda \cong \bigoplus_{\mu \in \mathbb{Y}_{n-1} : \mu \nearrow \lambda} \pi_\mu, \qquad \lambda \in \mathbb{Y}_n, \quad (1.4)$$

which plays a key role in an inductive argument to show the following property.

Proposition 1.3 *The set of $\{\pi_\lambda\}_{\lambda \in \mathbb{Y}_n}$ forms a complete system of representatives of the equivalence classes of irreducible representations of $\mathfrak{S}_n$.*

Hence (1.4) implies a multiplicity-free irreducible decomposition. Essential parts of the proofs omitted above are covered by a relation between Specht polynomials called the Garnir relation. My favorite textbook for the account is [27]. An alternative approach due to Okounkov–Vershik is contained in [6].

Symmetric Functions

Let Λ_n^k be the set of homogeneous symmetric polynomials of degree k in n variables, which contains for example

- monomial : $\lambda \in \mathbb{Y}_k$,

$$m_\lambda(x_1, \ldots, x_n) = \sum_{(\alpha_1, \ldots, \alpha_n)} x_1^{\alpha_1} \ldots x_n^{\alpha_n}$$

$((\alpha_1, \ldots, \alpha_n)$ runs over all distinct permutations of $(\lambda_1, \ldots, \lambda_{l(\lambda)}, 0, \ldots, 0))$,
- power sum :
$$p_k(x_1, \ldots, x_n) = x_1^k + \cdots + x_n^k,$$

- Schur polynomial : $\lambda \in \mathbb{Y}_k$, $l(\lambda) \leqq n$,

$$s_\lambda(x_1, \ldots, x_n) = \det[x_i^{\lambda_j + n - j}] / \det[x_i^{n-j}],$$

- complete symmetric polynomial:

$$h_k(x_1, \ldots, x_n) = \sum_{\lambda \in \mathbb{Y}_k} m_\lambda(x_1, \ldots, x_n).$$

Along the projective system $p_{nm} : \Lambda_m^k \longrightarrow \Lambda_n^k$, $n < m$, sending $f(x_1, \ldots, x_m)$ to $p_{nm} f = f(x_1, \ldots, x_n, 0, \ldots, 0)$, let Λ^k be the projective limit as $n \to \infty$. Then, m_λ ($\lambda \in \mathbb{Y}_k$), p_k and h_k are readily defined as elements of Λ^k. It is convenient to use the notation of a formal power series like $p_k = x_1^k + x_2^k + \cdots$. For Schur polynomials also, since we have for $\lambda \in \mathbb{Y}_k$

$$\begin{cases} s_\lambda(x_1, \ldots, x_n, 0) = s_\lambda(x_1, \ldots, x_n), & l(\lambda) \leqq n, \\ s_\lambda(x_1, \ldots, x_n, 0) = 0, & l(\lambda) = n + 1, \end{cases}$$

s_λ is well-defined as an element of Λ^k. An element of $\Lambda = \bigoplus_{k=0}^\infty \Lambda^k$ is called a symmetric function. The totality of Young diagrams of arbitrary sizes is denoted by $\mathbb{Y} = \bigsqcup_{k=0}^\infty \mathbb{Y}_k$. Here $\mathbb{Y}_0 = \{\varnothing\}$ is a singleton set. Now we have monomial symmetric function m_λ and Schur function s_λ for $\lambda \in \mathbb{Y}$. As power sum symmetric function p_λ and complete symmetric function h_λ for $\lambda \in \mathbb{Y}$, we set

$$p_\lambda = p_{\lambda_1} \cdots p_{\lambda_{l(\lambda)}}, \qquad h_\lambda = h_{\lambda_1} \ldots h_{\lambda_{l(\lambda)}},$$

furthermore $m_\varnothing = p_\varnothing = h_\varnothing = 1$.

Proposition 1.4 *Either* $\{m_\lambda\}_{\lambda \in \mathbb{Y}}$, $\{p_\lambda\}_{\lambda \in \mathbb{Y}}$ *or* $\{h_\lambda\}_{\lambda \in \mathbb{Y}}$ *forms a basis of* Λ.

Characters of $\mathfrak{S}_n$

Let χ^λ denote the character of an irreducible representation of $\mathfrak{S}_n$ corresponding to $\lambda \in \mathbb{Y}_n$, $\tilde{\chi}^\lambda$ be the normalized one, and χ_ρ^λ denote the value at $x \in C_\rho$ (= conjugacy class of cycle type $\rho \in \mathbb{Y}_n$):

$$\chi_\rho^\lambda = \chi^\lambda(x) = \operatorname{tr} \pi_\lambda(x), \qquad \tilde{\chi}_\rho^\lambda = \chi_\rho^\lambda / \dim \lambda.$$

There exists a bijective correspondence between $\mathcal{K}(\mathfrak{S}_n)$, the set of positive-definite, central, normalized complex-valued functions on $\mathfrak{S}_n$, and $\mathcal{P}(\mathbb{Y}_n)$, the set of probabilities on $\mathbb{Y}_n$, as $f \in \mathcal{K}(\mathfrak{S}_n) \longleftrightarrow \mathbb{M} \in \mathcal{P}(\mathbb{Y}_n)$:

$$f = \sum_{\lambda \in \mathbb{Y}_n} \mathbb{M}(\{\lambda\}) \tilde{\chi}^\lambda. \tag{1.5}$$

Proposition 1.5 (The Frobenius character formula I) *For* $k, n \in \mathbb{N}$ *and* $\rho \in \mathbb{Y}_n$,

$$p_\rho(x_1, \ldots, x_k) = \sum_{\lambda \in \mathbb{Y}_n : l(\lambda) \leqq k} \chi_\rho^\lambda s_\lambda(x_1, \ldots, x_k). \tag{1.6}$$

A fantastic way for showing (1.6) is to consider actions of the symmetric group $\mathfrak{S}_n$ and the unitary group $U(k)$ onto $(\mathbb{C}^k)^{\otimes n}$ and to apply the Schur–Weyl duality. Passing from (1.6) to the symmetric function setting yields the following.

Theorem 1.1 (The Frobenius character formula II) *For $n \in \mathbb{N}$ and $\rho, \lambda \in \mathbb{Y}_n$,*

$$p_\rho = \sum_{\lambda \in \mathbb{Y}_n} \chi_\rho^\lambda s_\lambda, \qquad s_\lambda = \sum_{\rho \in \mathbb{Y}_n} \frac{1}{z_\rho} \chi_\rho^\lambda p_\rho. \tag{1.7}$$

Note that the two expressions in (1.7) are connected by the orthogonality relation for the irreducible characters χ_ρ^λ.

The formula giving the value of χ^λ at a cycle is also well-known. We often use the notation $(k, 1^{n-k})$ instead of $(1^{n-k}k^1) = (k, 1, \dots, 1) \in \mathbb{Y}_n$. The descending kth power $z(z-1)\dots(z-k+1)$ is written simply as $z^{\downarrow k}$. The notation $[z^{-1}]\{\dots\}$ means the coefficient of z^{-1}-term in the Laurent series $\{\dots\}$.

Theorem 1.2 *For $n \in \mathbb{N}$, $k \in \{1, \dots, n\}$ and $\lambda \in \mathbb{Y}_n$,*

$$n^{\downarrow k} \tilde{\chi}_{(k,1^{n-k})}^\lambda = -\frac{1}{k}[z^{-1}]\left\{ z^{\downarrow k} \prod_{j=1}^n \frac{z - k - (\lambda_j + n - j)}{z - (\lambda_j + n - j)} \right\}. \tag{1.8}$$

We refer to [22] for getting informations on the symmetric functions and the characters of $\mathfrak{S}_n$. Also recommended for the same purpose is [24] which contains clear expositions.

1.2 Young Graph

In this section we recall basic notions on the Young graph and the infinite symmetric group and recognize the fundamental correspondence (1.13) of the three objects. The graph consisting of the vertex set $\mathbb{Y}$ and the edge structure defined by $\mu \nearrow \lambda$ in (1.4) is called the Young graph, which grows as seen in Fig. 1.3.

Harmonic Functions

If restriction is switched to induction, (1.4) is rephrased as

$$\mathrm{Ind}_{\mathfrak{S}_{n-1}}^{\mathfrak{S}_n} \pi_\lambda \cong \bigoplus_{\mu \in \mathbb{Y}_n : \lambda \nearrow \mu} \pi_\mu, \qquad \lambda \in \mathbb{Y}_{n-1}. \tag{1.9}$$

A complex-valued function φ on $\mathbb{Y}$ is said to be harmonic if

$$\varphi(\lambda) = \sum_{\mu \in \mathbb{Y} : \lambda \nearrow \mu} \varphi(\mu), \qquad \lambda \in \mathbb{Y},$$

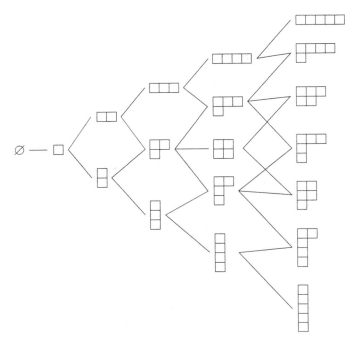

Fig. 1.3 Young graph

and normalized if $\varphi(\varnothing) = 1$. Let $\mathscr{H}(\mathbb{Y})$ denote the set of nonnegative normalized harmonic functions on $\mathbb{Y}$. Equip $\mathscr{H}(\mathbb{Y})$ with the topology of pointwise convergence of functions on $\mathbb{Y}$. Then, $\mathscr{H}(\mathbb{Y})$ is convex, compact and metrizable. Furthermore, $\mathscr{H}(\mathbb{Y})$ has a bijective correspondence to

$$\left\{ \psi : \Lambda \longrightarrow \mathbb{C} \,\middle|\, \text{linear}, \ \psi(1) = 1, \ \psi(s_\lambda) \geqq 0, \ \ker\psi \supset (s_1 - 1)\Lambda \right\}$$

by

$$\varphi(\lambda) = \psi(s_\lambda), \qquad \lambda \in \mathbb{Y}. \tag{1.10}$$

Indeed, harmonicity of φ is connected to the Pieri formula for Schur functions s_λ.

Central Probabilities

Let $\mathfrak{T}$ denote the set of infinite paths on the Young graph beginning at $\varnothing$. A path $t \in \mathfrak{T}$ is expressed as $t = \big(t(0) \nearrow t(1) \nearrow t(2) \nearrow \cdots\big)$ where $t(n) \in \mathbb{Y}_n$. The set of finite paths terminating at $\lambda \in \mathbb{Y}$ is denoted by $\mathfrak{T}(\lambda)$. Thus $\mathfrak{T}_n = \bigsqcup_{\lambda \in \mathbb{Y}_n} \mathfrak{T}(\lambda)$ is the set of paths of length n. Equip $\mathfrak{T}$ with the canonical projective limit topology induced by $t \in \mathfrak{T} \mapsto t_n \in \mathfrak{T}_n$, and $\mathfrak{T}$ is compact. A permutation σ of $\mathfrak{T}(\lambda)$, $\lambda \in \mathbb{Y}_n$, acts on $\mathfrak{T}$: $\sigma(t) = \big(\sigma(t_n) \nearrow t(n+1) \nearrow t(n+2) \nearrow \cdots\big)$ if $t \in \mathfrak{T}$ passes through λ, or $\sigma(t) = t$ otherwise. Let $\mathfrak{S}(\lambda)$ be all such transformations on $\mathfrak{T}$. The transformation group of $\mathfrak{T}$ generated by $\bigcup_{\lambda \in \mathbb{Y}} \mathfrak{S}(\lambda)$ is denoted by $\mathfrak{S}_0(\mathbb{Y})$. The Borel field of $\mathfrak{T}$, denoted by

$\mathscr{B}(\mathfrak{T})$, is generated by cylindrical subsets $C_u \subset \mathfrak{T}$ where $C_u = \{t \in \mathfrak{T} \mid t_n = u\}$ for $u \in \mathfrak{T}_n$. Let $\mathscr{P}(\mathfrak{T})$ denote the set of probabilities on $(\mathfrak{T}, \mathscr{B}(\mathfrak{T}))$. An element $M \in \mathscr{P}(\mathfrak{T})$ is $\mathfrak{S}_0(\mathbb{Y})$-invariant if and only if $M(C_u) = M(C_v)$ holds whenever $u(n) = v(n)$ for any $n \in \mathbb{N}$ and $u, v \in \mathfrak{T}_n$. We refer to an $\mathfrak{S}_0(\mathbb{Y})$-invariant probability as a central probability on $\mathfrak{T}$. Let $\mathscr{M}(\mathfrak{T})$ denote the set of central probabilities on $\mathfrak{T}$, and $\mathscr{M}(\mathfrak{T})$ is closed with respect to the weak convergence topology on $\mathscr{P}(\mathfrak{T})$ hence a compact set.

Lemma 1.1 *There exists an affine homeomorphism between the two compact convex sets $\mathscr{H}(\mathbb{Y}) \cong \mathscr{M}(\mathfrak{T})$ by*

$$\varphi(\lambda) = M(C_u), \qquad \lambda = u(n), \ \lambda \in \mathbb{Y}_n, \ u \in \mathfrak{T}_n. \tag{1.11}$$

The Infinite Symmetric Group

The infinite symmetric group $\mathfrak{S}_\infty$ is the inductive limit of (1.1), or, regarding an element of $\mathfrak{S}_n$ as a permutation of $\mathbb{N}$, $\mathfrak{S}_\infty = \bigcup_{n=1}^\infty \mathfrak{S}_n$. The identity element of $\mathfrak{S}_\infty$ is denoted by e. The support of $x \in \mathfrak{S}_\infty$, denoted by $\operatorname{supp} x$, is well-defined from those in $\mathfrak{S}_n$. A complex-valued function f on $\mathfrak{S}_\infty$ is said to be positive-definite if $\sum_{j,k=1}^l \overline{\alpha_j} \alpha_k f(x_j^{-1} x_k) \geqq 0$ for any $l \in \mathbb{N}$ and $x_j \in \mathfrak{S}_\infty, \alpha_j \in \mathbb{C}$ ($j \in \{1, \ldots, l\}$) and normalized if $f(e) = 1$. Let $\mathscr{K}(\mathfrak{S}_\infty)$ be the set of positive-definite, normalized and central complex-valued functions on $\mathfrak{S}_\infty$. Equip $\mathscr{K}(\mathfrak{S}_\infty)$ with the topology of pointwise convergence, and $\mathscr{K}(\mathfrak{S}_\infty)$ is compact, convex and metrizable.

Lemma 1.2 *There exists an affine homeomorphism $\mathscr{K}(\mathfrak{S}_\infty) \cong \mathscr{H}(\mathbb{Y})$ by*

$$f\big|_{\mathfrak{S}_n} = \sum_{\lambda \in \mathbb{Y}_n} \varphi(\lambda) \chi^\lambda, \qquad n \in \mathbb{N}. \tag{1.12}$$

Combining Lemmas 1.1 and 1.2, we have affine homeomorphisms

$$\mathscr{K}(\mathfrak{S}_\infty) \cong \mathscr{H}(\mathbb{Y}) \cong \mathscr{M}(\mathfrak{T}) \tag{1.13}$$

in which the mutual correspondences between $f \in \mathscr{K}(\mathfrak{S}_\infty)$, $\varphi \in \mathscr{H}(\mathbb{Y})$ and $M \in \mathscr{M}(\mathfrak{T})$ are given by (1.12) and (1.11).

The conjugacy classes of $\mathfrak{S}_\infty$ are parametrized by

$$\mathbb{Y}^\times = \{\rho \in \mathbb{Y} \mid m_1(\rho) = 0\}$$

where $\rho \in \mathbb{Y}^\times$ indicates the cycle type of nontrivial cycles of $x \in \mathfrak{S}_\infty$.

Extremal Objects

Since (1.13) is affine homeomorphisms between compact, convex and metrizable spaces, the subspaces consisting of the extremal points are also preserved under (1.13). Customarily, an extremal element of $\mathscr{K}(\mathfrak{S}_\infty)$, $\mathscr{H}(\mathbb{Y})$ and $\mathscr{M}(\mathfrak{T})$ is respectively called a character, a minimal harmonic function and an ergodic probability.

Theorem 1.3 (Thoma [28]) *An element $f \in \mathcal{K}(\mathfrak{S}_\infty)$ is a character of $\mathfrak{S}_\infty$ if and only if it is multiplicative, that is, $f(xy) = f(x)f(y)$ holds for $x, y \in \mathfrak{S}_\infty \setminus \{e\}$ such that $\operatorname{supp} x \cap \operatorname{supp} y = \varnothing$.*

Concerning the correspondence of (1.10) for $\mathcal{H}(\mathbb{Y})$, the following holds.

Proposition 1.6 *Under (1.10), $\varphi \in \mathcal{H}(\mathbb{Y})$ is extremal if and only if ψ is an algebra homomorphism.*

The extremal points of these spaces are parametrized by the well-known Thoma simplex. We call the subset of $[0, 1]^\infty \times [0, 1]^\infty$:

$$\Delta = \Big\{ (\alpha, \beta) \,\Big|\, \alpha = (\alpha_i)_{i=1}^\infty, \ \beta = (\beta_i)_{i=1}^\infty, \ \alpha_1 \geqq \alpha_2 \geqq \cdots \geqq 0, \ \beta_1 \geqq \beta_2 \geqq \cdots \geqq 0,$$
$$\sum_{i=1}^\infty (\alpha_i + \beta_i) \leqq 1 \Big\} \tag{1.14}$$

the Thoma simplex. Equipped with the relative topology of $[0, 1]^\infty \times [0, 1]^\infty$ (with the product topology), Δ is compact and metrizable.

Theorem 1.4 (Thoma [28]) *The set of characters of $\mathfrak{S}_\infty$ is homeomorphic to Δ. The correspondence $(\alpha, \beta) \in \Delta \leftrightarrow f$ (extremal in $\mathcal{K}(\mathfrak{S}_\infty)$) is given by*

$$f(k\text{-cycle}) = \sum_{i=1}^\infty \big(\alpha_i^k + (-1)^{k-1} \beta_i^k \big), \qquad k \in \{2, 3, \ldots\}. \tag{1.15}$$

Theorem 1.3 yields that (1.15) completely determines the values of character f. Furthermore, it is known that any element of $\mathcal{K}(\mathfrak{S}_\infty)$ has an integral representation over Δ and hence there exists an affine homeomorphism

$$\mathcal{K}(\mathfrak{S}_\infty) \cong \mathcal{P}(\Delta). \tag{1.16}$$

This is a variant of the classical Bochner theorem. By virtue of (1.13), Theorems 1.4 and (1.16) are translated into both $\mathcal{H}(\mathbb{Y})$ and $\mathcal{M}(\mathfrak{T})$.

The most fundamental extremal object is the one corresponding to $(\alpha, \beta) = (0, 0) \in \Delta$ in (1.14). In terms of a character of $\mathfrak{S}_\infty$, this agrees with $f_{0,0} = \delta_e$, the delta function at $e \in \mathfrak{S}_\infty$. Translating it into $\mathcal{M}(\mathfrak{T})$, we obtain the Plancherel measure M_{Pl} on $\mathfrak{T}$: for $n \in \mathbb{N}$,

$$M_{\mathrm{Pl}}(C_u) = \frac{\dim \lambda}{n!}, \qquad u \in \mathfrak{T}_n, \quad u(n) = \lambda \in \mathbb{Y}_n. \tag{1.17}$$

The Plancherel measure is thus an ergodic probability on $\mathfrak{T}$. The nth marginal distribution of M_{Pl}:

$$M_{\mathrm{Pl}}^{(n)}(\{\lambda\}) = M_{\mathrm{Pl}}(\{t \in \mathfrak{T} \mid t(n) = \lambda\}) = (\dim \lambda)^2 / n!, \qquad \lambda \in \mathbb{Y}_n \tag{1.18}$$

is also called the Plancherel measure on $\mathbb{Y}_n$.

All the materials presented in this section are well-known, but included in [13] with full proofs.

1.3 Free Probability

The readers who are not familiar with free probability and feel its appearance here a bit sudden may temporarily skip this section and revisit it after recognizing the necessity of relevant notions.

Cumulant

The kth (classical) cumulant $C_k(\mu)$ of $\mu \in \mathscr{P}(\mathbb{R})$ appears by definition in the coefficient of ζ^k in the expansion of logarithm of the Laplace transform of μ (with an appropriate exponential integrability condition):

$$\log \int_{\mathbb{R}} e^{\zeta x} \mu(dx) = \sum_{k=1}^{\infty} \frac{C_k(\mu)}{k!} \zeta^k.$$

In other words, using the nth moment of μ: $M_n(\mu) = \int_{\mathbb{R}} x^n \mu(dx)$, we have

$$\sum_{n=0}^{\infty} \frac{M_n(\mu)}{n!} \zeta^n = \exp\Big(\sum_{k=1}^{\infty} \frac{C_k(\mu)}{k!} \zeta^k\Big). \tag{1.19}$$

Comparing the coefficients of both sides of (1.19), we obtain the cumulant-moment formula as follows. Let $P(n)$ denote the set of partitions of $\{1, 2, \ldots, n\}$. By definition $\pi = \{v_1, \ldots, v_l\} \in P(n)$, $v_i \neq \varnothing$, gives $\{1, 2, \ldots, n\} = v_1 \sqcup \cdots \sqcup v_l$, where each v_i is called a block of π and $l = b(\pi)$ denotes the number of blocks of π. For $\pi, \rho \in P(n)$, if any block of ρ is a subset of some block of π, we write as $\rho \leq \pi$. Clearly, $P(n)$ is a poset with the minimal element $0_n = \{\{1\}, \{2\}, \ldots, \{n\}\}$ and the maximal element $1_n = \{\{1, 2, \ldots, n\}\}$. Cumulants of μ are extended to the partition subscript case in a multiplicative way:

$$C_\pi(\mu) = \prod_{i=1}^{b(\pi)} C_{|v_i|}(\mu), \qquad \pi = \{v_1, \ldots, v_{b(\pi)}\} \in P(n) \tag{1.20}$$

where $|v_i|$ is the cardinality of block v_i. Then, (1.19) yields the following.

Proposition 1.7 *For* $\mu \in \mathscr{P}(\mathbb{R})$,

$$M_n(\mu) = \sum_{\pi \in P(n)} C_\pi(\mu), \qquad n \in \mathbb{N}. \tag{1.21}$$

Moments of μ are also extended multiplicatively with respect to the blocks as (1.20). By using the Möbius function $m_{P(n)}$ for poset $P(n)$, we can invert (1.21).

Proposition 1.8 *For* $\mu \in \mathscr{P}(\mathbb{R})$,

$$C_n(\mu) = \sum_{\pi \in P(n)} m_{P(n)}(\pi, 1_n) M_\pi(\mu), \qquad n \in \mathbb{N}. \qquad (1.22)$$

Here $m_{P(n)}(\rho, \pi)$ is determined as the inverse matrix of $a_{P(n)}(\rho, \pi) = 1_{\{\rho \leq \pi\}}$. Note that (1.21) and (1.22) (both with multiplicative extensions), called cumulant-moment formulas, serve as a definition of the cumulant $C_k(\mu)$ for any $\mu \in \mathscr{P}(\mathbb{R})$ having all moments.

A partition $\pi \in P(n)$ is often described by connecting all the letters in a block by an arc as indicated in Fig. 1.4. We call π a non-crossing partition if it is expressed with no crossing arcs in such a description. In Fig. 1.4, the 14 partitions (except the 13th one) are non-crossing. A non-crossing partition is called an interval partition if no arcs are nested. In Fig. 1.4, the first and second are interval partitions, while the third and fourth are not. The posets of non-crossing partitions and interval partitions of $\{1, 2, \ldots, n\}$ are denoted by $NC(n)$ and $I(n)$ respectively. We thus have $I(n) \subset NC(n) \subset P(n)$. Replacing $P(n)$ by $NC(n)$, we introduce the kth free cumulant $R_k(\mu)$ for $\mu \in \mathscr{P}(\mathbb{R})$. The free cumulant-moment formulas then take the following forms.

Proposition 1.9 *For* $\mu \in \mathscr{P}(\mathbb{R})$ *and* $n \in \mathbb{N}$,

$$M_n(\mu) = \sum_{\pi \in NC(n)} R_\pi(\mu), \qquad R_n(\mu) = \sum_{\pi \in NC(n)} m_{NC(n)}(\pi, 1_n) M_\pi(\mu). \qquad (1.23)$$

Here $m_{NC(n)}$ *is the Möbius function for poset* $NC(n)$.

Moreover, adopting also $I(n)$ as a partition structure, we obtain Boolean cumulants $B_k(\mu)$ for $\mu \in \mathscr{P}(\mathbb{R})$ and the Boolean cumulant-moment formulas similar to (1.21), (1.22) and (1.23).

Convolution

The convolution $\mu * \nu$ of $\mu, \nu \in \mathscr{P}(\mathbb{R})$ is linearized by the cumulants:

$$C_k(\mu * \nu) = C_k(\mu) + C_k(\nu), \qquad k \in \mathbb{N}.$$

Analogously, we introduce the free convolution $\mu \boxplus \nu$ of $\mu, \nu \in \mathscr{P}(\mathbb{R})$ which satisfies

$$R_k(\mu \boxplus \nu) = R_k(\mu) + R_k(\nu), \qquad k \in \mathbb{N}. \qquad (1.24)$$

Fig. 1.4 $|P(4)| = 15$, $|NC(4)| = 14$, $|I(4)| = 8$

Equivalently, in terms of the free cumulant-moment formula (1.23), $\mu \boxplus \nu$ is a probability on $\mathbb{R}$ whose moments are given by

$$M_n(\mu \boxplus \nu) = \sum_{\pi = \{\nu_1, \cdots, \nu_{b(\pi)}\} \in NC(n)} \prod_{i=1}^{b(\pi)} \left(R_{|\nu_i|}(\mu) + R_{|\nu_i|}(\nu) \right), \qquad n \in \mathbb{N}. \quad (1.25)$$

Some extra conditions for μ and ν are needed in addition to the existence of all moments, in order for (1.25) to determine $\mu \boxplus \nu$ uniquely. There are no problems if μ and ν have compact supports, and then so does $\mu \boxplus \nu$.

Generating Function

At a level of (exponential) generating functions, the moments and cumulants of $\mu \in \mathscr{P}(\mathbb{R})$ are connected to each other by (1.19). For a free cumulant sequence $\{R_k(\mu)\}_{k \in \mathbb{N}}$, we consider (as formal series)

$$R_\mu(\zeta) = \sum_{k=0}^{\infty} R_{k+1}(\mu) \zeta^k, \qquad K_\mu(\zeta) = \frac{1}{\zeta} + R_\mu(\zeta). \quad (1.26)$$

We call $R_\mu(\zeta)$ Voiculescu's R-transform of μ. The Stieltjes transform

$$G_\mu(z) = \int_{\mathbb{R}} \frac{1}{z - x} \mu(dx) = \sum_{n=0}^{\infty} \frac{M_n(\mu)}{z^{n+1}}$$

of μ is another generating function of the moments of μ. The free cumulant-moment formula (1.23) is now converted into the following form.

Proposition 1.10 *If $\mu \in \mathscr{P}(\mathbb{R})$ has a compact support, there exists $\delta > 0$ such that $K_\mu(\zeta)$ is holomorphic in $0 < |\zeta| < \delta$ and yields $K_\mu(\zeta) = G_\mu^{-1}(\zeta)$.*

A generating function of the Boolean cumulants of $\mu \in \mathscr{P}(\mathbb{R})$ is derived in a similar (in fact, easier) way to Proposition 1.10. We will recall it in introducing the Kerov polynomials (Theorem 2.2).

Proposition 1.11 *If $\mu \in \mathscr{P}(\mathbb{R})$ has a compact support, $G_\mu(z)^{-1}$ is holomorphic in a large annulus $a < |z| < \infty$ with the Laurent expansion:*

$$\frac{1}{G_\mu(z)} = z - \sum_{k=1}^{\infty} \frac{B_k(\mu)}{z^{k-1}}. \quad (1.27)$$

Proof Since $G_\mu(z)^{-1}$ is holomorphic in $|z| \gg 1$ and satisfies $\lim_{z \to \infty} z G_\mu(z) = 1$, it has the Laurent expansion:

$$G_\mu(z)^{-1} = z + \sum_{k=1}^{\infty} \frac{c_k}{z^{k-1}}, \qquad |z| \gg 1. \quad (1.28)$$

Lemma 1.3 below with $\alpha_n = M_n(\mu)$ and $\gamma_k = B_k(\mu)$ yields

$$G\left(\frac{1}{\zeta}\right) = \zeta A(\zeta), \qquad C(\zeta) = \sum_{k=1}^{\infty} B_k \zeta^k$$

and (1.30). Therefore, comparing $G(1/\zeta)^{-1} = \zeta^{-1} + \sum_{k=1}^{\infty} c_k \zeta^{k-1}$ ($|\zeta| \ll 1$) obtained by (1.28) to

$$G\left(\frac{1}{\zeta}\right)^{-1} = \frac{1}{\zeta A(\zeta)} = \frac{1}{\zeta}(1 - C(\zeta)) = \frac{1}{\zeta} - \sum_{k=1}^{\infty} B_k \zeta^{k-1},$$

we have $c_k = -B_k$ for any $k \in \mathbb{N}$. This completes the proof of (1.27).

Lemma 1.3 *Given real sequences $\{\alpha_n\}_{n \in \mathbb{N}}$ and $\{\gamma_k\}_{k \in \mathbb{N}}$, consider formal power series*

$$A(\zeta) = 1 + \sum_{n=1}^{\infty} \alpha_n \zeta^n, \qquad C(\zeta) = \sum_{k=1}^{\infty} \gamma_k \zeta^k$$

and define γ_π multiplicatively for $\pi \in I(n)$ from γ_k's. Then the following are equivalent:

$$\bullet \; \alpha_n = \sum_{\pi \in I(n)} \gamma_\pi, \qquad n \in \mathbb{N}, \tag{1.29}$$

$$\bullet \; A(\zeta)C(\zeta) = A(\zeta) - 1. \tag{1.30}$$

Proof We rewrite (1.30) as the relation between the coefficients:

$$\alpha_n = \sum_{l=1}^{n} \alpha_{n-l} \gamma_l, \qquad n \in \mathbb{N} \tag{1.31}$$

with setting $\alpha_0 = 1$. It suffices to verify that α_n's determined by (1.29) satisfy the recurrence (1.31). Dividing the interval partitions according to length of the block containing 1, we have

$$\alpha_n = \sum_{\pi \in I(n)} \gamma_\pi = \gamma_n + \sum_{l=1}^{n-1} \sum_{\rho \in I(n-l)} \gamma_l \gamma_\rho = \sum_{l=1}^{n} \gamma_l \alpha_{n-l}$$

as desired.

Proposition 1.12 *If $\mu \in \mathscr{P}(\mathbb{R})$ has a compact support, the free cumulants are expressed as*

$$R_k(\mu) = -\frac{1}{2\pi(k-1)i} \int_{\{|z|=s\}} \frac{dz}{G_\mu(z)^{k-1}} = -\frac{1}{k-1}[z^{-1}]\left(\frac{1}{G_\mu(z)^{k-1}}\right) \tag{1.32}$$

for $k \in \{2, 3, \ldots\}$ and sufficiently large $s > 0$.

Proof Noting $G_\mu(z)$ and $G_\mu(z)^{-1}$ are holomorphic in $|z| \gg 1$, we put $\zeta = G_\mu(z)$ in the integral expression for $R_k(\mu)$ induced from (1.26):

$$R_k(\mu) = \frac{1}{2\pi i} \int_{\{|\zeta|=r\}} \frac{K_\mu(\zeta)}{\zeta^k} d\zeta = \frac{1}{2\pi i} \int_{-\{|z|=s\}} \frac{z}{G_\mu(z)^k} G'_\mu(z) dz$$

$$= -\frac{1}{2\pi(k-1)i} \int_{\{|z|=s\}} \frac{dz}{G_\mu(z)^{k-1}}.$$

Note that, if ζ runs over $\{|\zeta| = r\}$ in the ordinary direction, z runs over a simple closed curve lying in an annulus large enough in the reverse direction.

Freeness

If $\mathbb{R}$-valued independent random variables a and b have distributions μ and ν respectively, the distribution of $a + b$ is given by their convolution $\mu * \nu$. On the other hand, the free convolution comes from the important notion of freeness of noncommutative random variables. A pair (A, ϕ) of unital $*$-algebra A (over $\mathbb{C}$) and state ϕ of A is called a probability space. A family $\{A_\alpha\}$ of unital $*$-subalgebras of A are said to be free in (A, ϕ), or with respect to ϕ, if the following are fulfilled: for any $n \in \mathbb{N}$,

$$\begin{cases} a_i \in A_{\alpha_i}, & i \in \{1, \ldots, n\} \\ \phi(a_i) = 0, & i \in \{1, \ldots, n\} \implies \phi(a_1 a_2 \ldots a_n) = 0 \\ \alpha_1 \neq \alpha_2 \neq \cdots \neq \alpha_n \end{cases}$$

(the last assumption means that any adjacent α_i's are distinct). Two random variables $a, b \in A$ are said to be free if the generated $*$-subalgebras $\langle a, a^* \rangle$ and $\langle b, b^* \rangle$ are free. For self-adjoint $a \in A$ and $\mu \in \mathscr{P}(\mathbb{R})$, we say a obeys μ, or the distribution of a is μ, and write as $a \sim \mu$ if $\phi(a^n) = M_n(\mu)$ holds for any $n \in \mathbb{N}$ (admitting that the moment sequence $\{\phi(a^n)\}_{n \in \mathbb{N}}$ does not necessarily determine a unique probability on $\mathbb{R}$).

Proposition 1.13 *If $a, b \in A$ are free, $a \sim \mu$, $b \sim \nu$ and μ, ν have compact supports, then $a + b \sim \mu \boxplus \nu$.*

Let $q \in A$ be a projection, $q^2 = q = q^*$, such that $\phi(q) \neq 0$. Setting $B = qAq$ and $\psi = \phi(q)^{-1}\phi|_B$, we have a new probability space (B, ψ). If self-adjoint $a \in A$ and q are free, the distribution of qaq in (B, ψ) is called the free compression of μ, where $a \sim \mu \in \mathscr{P}(\mathbb{R})$. For compactly supported $\mu \in \mathscr{P}(\mathbb{R})$ and $0 < c \leq 1$, the free compression is uniquely determined and denoted by $\mu_c \in \mathscr{P}(\mathbb{R})$.

Proposition 1.14 *The free compression μ_c of $\mu \in \mathscr{P}(\mathbb{R})$ is characterized in terms of free cumulants by*

$$R_k(\mu_c) = c^{k-1} R_k(\mu), \quad k \in \mathbb{N}. \tag{1.33}$$

Readers should consult [32] above all to know what free probability means. All informations on free probability theory needed for our purpose are contained in [23].

Chapter 2
Analysis of the Kerov–Olshanski Algebra

Abstract In this chapter, we investigate the algebra of polynomial functions in coordinates of Young diagrams as a nice framework in which various quantities on Young diagrams can be efficiently computed. This algebra was introduced by Kerov–Olshanski [20], analysis of which is substantially due to Ivanov–Olshanski [16]. Several systems of generators and associated generating functions are considered. It is important to understand the concrete transition rules between these generating systems, one of which is the Kerov polynomial.

2.1 Coordinates of a Young Diagram

In this section, we consider two kinds of coordinates encoding a Young diagram: the Frobenius coordinates and the min-max coordinates.

Let $\lambda = (\lambda_1 \geqq \lambda_2 \geqq \cdots) \in \mathbb{Y}$ be a Young diagram having d boxes along the main diagonal. We call

$$a_i = a_i(\lambda) = \lambda_i - i + \frac{1}{2}, \quad b_i = b_i(\lambda) = \lambda'_i - i + \frac{1}{2}, \quad i \in \{1, 2, \ldots, d\}$$

the Frobenius coordinates of λ and write as $\lambda = (a_1, \ldots, a_d \,|\, b_1, \ldots, b_d)$. The Frobenius coordinates of $\lambda \in \mathbb{Y}$ satisfy

$$-b_1 < -b_2 < \cdots < -b_d < 0 < a_d < \cdots < a_2 < a_1, \quad |\lambda| = \sum_{i=1}^{d}(a_i + b_i).$$

Let us display a Young diagram in the upper half of the xy-plane as in Fig. 2.1, where $\lambda = (4, 2, 2, 1)$ of the French style in Fig. 1.1 is rotated by $45°$ and put in such a way that the main diagonal boxes lie along the y-axis. The piecewise linear border indicated by bold lines in Fig. 2.1 is called the profile of a Young diagram. Since it is preferable that the corners of any profile have integral xy-coordinates, we always assume that the edge length of each box is $\sqrt{2}$ in the display as in Fig. 2.1.

© The Author(s) 2016
A. Hora, *The Limit Shape Problem for Ensembles of Young Diagrams*,
SpringerBriefs in Mathematical Physics, DOI 10.1007/978-4-431-56487-4_2

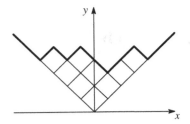 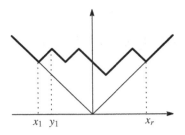

Fig. 2.1 *(left)* profile of $\lambda = (4, 2, 2, 1)$; *(right)* its min-max coordinates

For $\lambda = (\lambda_1 \geq \lambda_2 \geq \cdots) \in \mathbb{Y}$, the subset of $\mathbb{Z} + \frac{1}{2}$ defined by $M(\lambda) = \{\lambda_i - i + \frac{1}{2}\}_{i \in \mathbb{N}}$ is called the Maya diagram of λ. It is easy to see

$$\{a_1, \ldots, a_d\} = M(\lambda) \cap \left(\mathbb{N} - \frac{1}{2}\right), \qquad \{-b_1, \ldots, -b_d\} = \left(-M(\lambda')\right) \cap \left(-\mathbb{N} + \frac{1}{2}\right),$$

$$M(\lambda) \sqcup \left(-M(\lambda')\right) = \mathbb{Z} + \frac{1}{2}$$

for $\lambda = (a_1, \ldots, a_d \mid b_1, \ldots, b_d) \in \mathbb{Y}$. The set $\{b : \text{box} \mid b \in \lambda\}$ is bijective to $\left\{(s, t) \in M(\lambda) \times \left(-M(\lambda')\right) \mid s > t\right\}$. We have $h_\lambda(b) = s - t$ as the hook length under this bijective correspondence $b \leftrightarrow (s, t)$ and hence

$$\log \prod_{b \in \lambda} h_\lambda(b) = \sum_{(s,t) \in M(\lambda) \times (-M(\lambda')) : s > t} \log(s - t). \qquad (2.1)$$

Through the hook formula (Proposition 1.1) and (2.1), maximizing $\dim \lambda$ in $\mathbb{Y}_n$ is equivalent to minimizing the RHS of (2.1).

Given $\lambda = (a_1, \ldots, a_d \mid b_1, \ldots, b_d) \in \mathbb{Y}$, we consider a polynomial of degree k in the Frobenius coordinates:

$$p_k(\lambda) = \sum_{i=1}^{d} \left(a_i^k + (-1)^{k-1} b_i^k\right), \qquad k \in \mathbb{N}, \qquad (2.2)$$

and a rational function with a_i and $-b_i$ as its pole and zero respectively:

$$\Phi(z; \lambda) = \prod_{i=1}^{d} \frac{z + b_i}{z - a_i}, \qquad z \in \mathbb{C}. \qquad (2.3)$$

We may set $\Phi(z; \varnothing) = 1$ though we do not consider the Frobenius coordinates of the empty diagram $\varnothing$. In a sufficiently large annulus $1 \ll |z| < \infty$, the Laurent expansion of Φ gives

$$\Phi(z;\lambda) = \prod_{i=1}^{d} \frac{1 + (b_i/z)}{1 - (a_i/z)} = \exp\Big(\sum_{k=1}^{\infty} \frac{p_k(\lambda)}{k} z^{-k}\Big). \tag{2.4}$$

The x-coordinates of the interlacing valleys (=local minima) and peaks (=local maxima) of the profile of $\lambda \in \mathbb{Y}$ yields an integer sequence

$$x_1 < y_1 < x_2 < y_2 < \cdots < x_{r-1} < y_{r-1} < x_r, \qquad r \in \mathbb{N}, \tag{2.5}$$

which is called the min-max coordinates of λ. Clearly, the last x_r is determined from $x_1, \ldots, y_{r-1}$. It is not difficult to see the following characterization.

Lemma 2.1 *An interlacing real sequence of* (2.5) *forms the min-max coordinates of some* $\lambda \in \mathbb{Y}$ *if and only if*

$$\sum_{i=1}^{r} x_i - \sum_{i=1}^{r-1} y_i = 0 \quad and \quad x_1, \ldots, x_r, y_1, \ldots, y_{r-1} \in \mathbb{Z}.$$

We consider a rational function with min coordinate x_i and max coordinate y_i as its pole and zero respectively:

$$G(z;\lambda) = \frac{(z - y_1) \cdots (z - y_{r-1})}{(z - x_1) \cdots (z - x_r)}, \qquad z \in \mathbb{C}. \tag{2.6}$$

In particular, $G(z; \varnothing) = 1/z$ for the empty diagram.

Transposing λ to λ' in (2.3) and (2.6), we readily have

$$\Phi(z;\lambda') = \Phi(-z;\lambda)^{-1}, \quad G(z;\lambda') = -G(-z;\lambda), \qquad \lambda \in \mathbb{Y}, \ z \in \mathbb{C}.$$

Proposition 2.1 *The rational functions* Φ *of* (2.3) *and* G *of* (2.6) *for*

$$\lambda = (a_1, \ldots, a_d \mid b_1, \ldots, b_d) = (x_1 < y_1 < \cdots < y_{r-1} < x_r) \in \mathbb{Y}$$

are connected as

$$\frac{\Phi(z - \frac{1}{2}; \lambda)}{\Phi(z + \frac{1}{2}; \lambda)} = z\, G(z;\lambda), \qquad z \in \mathbb{C}. \tag{2.7}$$

Proof When we rewrite $\Phi(z;\lambda)$, which is expressed by the Frobenius coordinates of λ, in terms of the min-max coordinates, we have only to be careful about how the profile of λ traverses the y-axis. Consider the situations case by case.

2.2   Transition Measure I

In this section, we translate encoding of a Young diagram by its coordinates into two atomic measures on $\mathbb{R}$; one called Kerov's transition measure and the other the Rayleigh measure. Such embedding into the space of measures enables us to develop asymptotic theory in a flexible framework.

We begin with a bit wider class than Young diagrams. A function $\lambda : \mathbb{R} \longrightarrow \mathbb{R}$, or the graph $y = \lambda(x)$, satisfying the following conditions is called a (centered) rectangular diagram:

(i) continuous and piecewise linear   (ii) $\lambda'(x) = \pm 1$ except finite x's
(iii) $\lambda(x) = |x|$ for $|x|$ large enough.

The set of rectangular diagrams is denoted by $\mathbb{D}_0$. A rectangular diagram is (the profile of) a Young diagram if and only if the exceptional x's in (ii) are all integers. This yields the natural inclusion $\mathbb{Y} \subset \mathbb{D}_0$. The definitions of the min-max coordinates and the rational function G, (2.5) and (2.6) respectively, are immediately extended from $\mathbb{Y}$ to $\mathbb{D}_0$.

Lemma 2.2 *An interlacing real sequence of* (2.5) *forms the min-max coordinates of some* $\lambda \in \mathbb{D}_0$ *if and only if*

$$\sum_{i=1}^{r} x_i - \sum_{i=1}^{r-1} y_i = 0.$$

To $\lambda = (x_1 < y_1 < \cdots < y_{r-1} < x_r) \in \mathbb{D}_0$ we assign an $\mathbb{R}$-valued (probability) measure on $\mathbb{R}$ as

$$\tau_\lambda = \sum_{i=1}^{r} \delta_{x_i} - \sum_{i=1}^{r-1} \delta_{y_i} \tag{2.8}$$

and call it the Rayleigh measure of $\lambda \in \mathbb{D}_0$. Clearly, $\lambda \mapsto \tau_\lambda$ is injective. Under derivatives of Schwartz' distributions we have

$$\tau_\lambda = \left(\frac{\lambda(x) - |x|}{2} \right)'' + \delta_0. \tag{2.9}$$

Let us use the notation of the kth moment $M_k(\,\cdot\,)$ for an $\mathbb{R}$-valued measure on $\mathbb{R}$ also. Then (2.8) and (2.9) yield

$$M_k(\tau_\lambda) = \sum_{i=1}^{r} x_i^k - \sum_{i=1}^{r-1} y_i^k = \int_{\mathbb{R}} x^k \left(\frac{\lambda(x) - |x|}{2} \right)'' dx + \delta_{0k}, \qquad k \in \mathbb{N} \cup \{0\}. \tag{2.10}$$

In particular, we have through integration by parts

$$M_2(\tau_\lambda) = \int_{\mathbb{R}} \left(\lambda(x) - |x| \right) dx = 2|\lambda|. \tag{2.11}$$

Lemma 2.3 *We can reconstruct* $\lambda \in \mathbb{D}_0$ *from its Rayleigh measure* τ_λ *by*

$$\lambda(u) = \int_\mathbb{R} |u - x| \tau_\lambda(dx), \qquad u \in \mathbb{R}.$$

Proof We use (2.9), but note that $|u - x|$ is not differentiable. Take $a > 0$ such that $\text{supp}\,(\lambda(x) - |x|) \subset (-a, a)$. The function $(\lambda(x) - |x|)'$ is of bounded variation and $(\lambda(x) - |x|)''$ is an $\mathbb{R}$-valued measure, both supported in $(-a, a)$. For $u \in (-a, a)$

$$\int_{(-a,a)} |u - x| \left(\frac{\lambda(x) - |x|}{2}\right)'' dx$$
$$= \int_{(-a,u)} (u - x) \left(\frac{\lambda(x) - |x|}{2}\right)'' dx + \int_{(u,a)} (x - u) \left(\frac{\lambda(x) - |x|}{2}\right)'' dx. \quad (2.12)$$

The first term of the RHS of (2.12) is

$$\int_{(-a,u)} \left(\int_x^u dy\right) \left(\frac{\lambda(x) - |x|}{2}\right)'' dx = \int_{(-a,u)} \left(\int_{(-a,y)} \left(\frac{\lambda(x) - |x|}{2}\right)'' dx\right) dy$$
$$= \int_{(-a,u)} \left(\frac{\lambda(y) - |y|}{2}\right)' dy = \frac{\lambda(u) - |u|}{2},$$

and so is the second term. We thus have (2.12) to be $\lambda(u) - |u|$. The cases of $u \geq a$ and $u \leq a$ are easier to see. Combine this with $\int_{-\infty}^\infty |u - x| \delta_0(dx) = |u|$.

In order to define the transition measure of a rectangular diagram, we consider the partial fraction expansion of (2.6) for $\lambda = (x_1 < y_1 < \cdots < y_{r-1} < x_r) \in \mathbb{D}_0$:

$$G(z; \lambda) = \frac{(z - y_1) \cdots (z - y_{r-1})}{(z - x_1) \cdots (z - x_r)} = \frac{\mu_1}{z - x_1} + \cdots + \frac{\mu_r}{z - x_r}, \quad (2.13)$$

$$\mu_i = \frac{(x_i - y_1) \cdots (x_i - y_{r-1})}{(x_i - x_1) \cdots (x_i - x_{i-1})(x_i - x_{i+1}) \cdots (x_i - x_r)}, \quad i \in \{1, \ldots, r\}. \quad (2.14)$$

The interlacing property (2.5) assures $\mu_i > 0$ in (2.14). Multiplying (2.13) by z and letting $z \to \infty$ yield $\sum_{i=1}^r \mu_i = 1$. We thus have an atomic probability on $\mathbb{R}$

$$\mathfrak{m}_\lambda = \sum_{i=1}^r \mu_i \delta_{x_i} \in \mathscr{P}(\mathbb{R}), \qquad \text{supp}\,\mathfrak{m}_\lambda = \{x_1, \ldots, x_r\} \quad (2.15)$$

called (Kerov's) transition measure of $\lambda \in \mathbb{D}_0$. Note that (2.13) is the Stieltjes transform of $\mathfrak{m}_\lambda$:

$$G_{\mathfrak{m}_\lambda}(z) = \int_\mathbb{R} \frac{1}{z - x} \mathfrak{m}_\lambda(dx) = G(z; \lambda), \qquad z \in \mathbb{C}. \quad (2.16)$$

Proposition 2.2 *Given* $\lambda \in \mathbb{D}_0$, *the two moment sequences* $\{M_n(\mathfrak{m}_\lambda)\}_{n \in \mathbb{N}}$ *and* $\{M_k(\tau_\lambda)\}_{k \in \mathbb{N}}$ *are connected to each other by*

$$\sum_{n=0}^{\infty} M_n(\mathfrak{m}_\lambda) z^{-n} = \exp\Big(\sum_{k=1}^{\infty} \frac{M_k(\tau_\lambda)}{k} z^{-k} \Big). \tag{2.17}$$

Hence $\{M_n(\mathfrak{m}_\lambda)\}$ *and* $\{M_k(\tau_\lambda)\}$ *are expressed by polynomials in each other.*

Proof Setting $\mu = \sum_{i=1}^r \mu_i \delta_{x_i}$ in (2.13) for interlacing $x_1 < y_1 < \cdots < y_{r-1} < x_r$ and μ_i of (2.14), we get for $|z| \gg 1$

$$\sum_{n=0}^{\infty} M_n(\mu) z^{-n} = z\, G_\mu(z) = \frac{z(z - y_1) \cdots (z - y_{r-1})}{(z - x_1) \cdots (z - x_r)}$$

$$= \exp\Big\{ \sum_{i=1}^{r-1} \log\Big(1 - \frac{y_i}{z}\Big) - \sum_{i=1}^{r} \log\Big(1 - \frac{x_i}{z}\Big) \Big\} = \exp\Big\{ \sum_{k=1}^{\infty} \frac{1}{k} \Big(\sum_{i=1}^{r} x_i^k - \sum_{i=1}^{r-1} y_i^k \Big) z^{-k} \Big\}. \tag{2.18}$$

Specialization to the min-max coordinates of $\lambda \in \mathbb{D}_0$ yields (2.17).

As the terms of z^{-1} and z^{-2} in (2.17), we have

$$M_1(\mathfrak{m}_\lambda) = M_1(\tau_\lambda) = 0, \qquad M_2(\mathfrak{m}_\lambda) = \frac{1}{2} M_2(\tau_\lambda). \tag{2.19}$$

Proposition 2.3 *The map* $\lambda \mapsto \mathfrak{m}_\lambda$ *gives a bijection of* $\mathbb{D}_0$ *to the set of probabilities on* $\mathbb{R}$ *with mean* 0 *and finite supports.*

Proof Since the injectivity is immediate from (2.13), we verify the surjectivity. Take any

$$\mu = \sum_{i=1}^{r} \mu_i \delta_{x_i}, \qquad x_1 < \cdots < x_r, \qquad \mu_i > 0, \qquad \sum_{i=1}^{r} \mu_i = 1, \qquad \sum_{i=1}^{r} x_i \mu_i = 0.$$

Determine a monic real polynomial $f(z)$ of degree $r - 1$ by

$$\frac{\mu_1}{z - x_1} + \cdots + \frac{\mu_r}{z - x_r} = \frac{f(z)}{(z - x_1) \cdots (z - x_r)}.$$

Since $f(x_1), f(x_2), \ldots, f(x_r)$ have alternating sign changes, f has $r - 1$ zeros y_i satisfying $x_1 < y_1 < x_2 < \cdots < x_{r-1} < y_{r-1} < x_r$. We hence have the same equality as (2.13) and then (2.18), in particular $M_1(\mu) = \sum_{i=1}^{r} x_i - \sum_{i=1}^{r-1} y_i$ as the coefficient of z^{-1}. Lemma 2.2 assures the existence of $\lambda \in \mathbb{D}_0$ such that $\mathfrak{m}_\lambda = \mu$.

While the Rayleigh measure τ_λ reflects the shape of $\lambda \in \mathbb{Y}_n$ more or less directly, the transition measure $\mathfrak{m}_\lambda$ gives us information about the irreducible representation of $\mathfrak{S}_n$ labeled by λ. Let us see a few instances.

The Plancherel measure M_{Pl} on the path space $\mathfrak{T}$ defined by (1.17) induces a Markov chain on $\mathbb{Y}$. In fact, assuming $\lambda^0 = \varnothing \nearrow \lambda^1 \nearrow \cdots \nearrow \lambda^{n-1} \nearrow \lambda (\in \mathbb{Y}_n)$ forms a path in $\mathfrak{T}_n$, we have the conditional probability

$$M_{\mathrm{Pl}}\big(t(n+1) = \mu \,\big|\, t(0) = \lambda^0, \cdots, t(n) = \lambda\big)$$

$$= \begin{cases} \dfrac{M_{\mathrm{Pl}}(C_{\lambda^0 \nearrow \cdots \nearrow \lambda \nearrow \mu})}{M_{\mathrm{Pl}}(C_{\lambda^0 \nearrow \cdots \nearrow \lambda})} = \dfrac{\dim \mu}{(n+1)\dim \lambda}, & \lambda \nearrow \mu, \\[2mm] 0, & \text{otherwise.} \end{cases}$$

This chain is often called the Plancherel growth process. Let $(x_1 < y_1 < x_2 < \cdots < y_{r-1} < x_r)$ be the min-max coordinates of $\lambda \in \mathbb{Y}_n$ and $\mu^{(i)} \in \mathbb{Y}_{n+1}$ denote the Young diagram obtained by putting a box at the ith valley (of the x-coordinate x_i) of λ. The following fact gives a good reason for $\mathfrak{m}_\lambda$ to be called the transition measure.

Lemma 2.4 *Under the above notations,*

$$\mathfrak{m}_\lambda(\{x_i\}) = \frac{\dim \mu^{(i)}}{(n+1)\dim \lambda}, \qquad i \in \{1, \ldots, r\}. \tag{2.20}$$

Proof The hook formula (Proposition 1.1) implies that the RHS of (2.20) is

$$\prod_{b \in \lambda} h_\lambda(b) \Big/ \prod_{b \in \mu^{(i)}} h_{\mu^{(i)}}(b).$$

When we rewrite this quantity in terms of the min-max coordinates, we have only to focus on the boxes lying in zone I and zone II in Fig. 2.2, where $\mu^{(i)}/\lambda$ is the (p, q) box in $\mu^{(i)}$. The hook length at $(p, 1)$ box in zone I is $h_{\mu^{(i)}}(p, 1) = x_i - x_1$, and so on. Successive cancellations yield (2.14) and hence $\mathfrak{m}_\lambda(\{x_i\})$. $\qquad\blacksquare$

Fig. 2.2 Min-max coordinates and hook length ratio

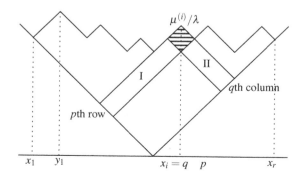

Theorem 1.2 tells the irreducible character value at a cycle, where (1.8) is expressed in terms of row lengths of a Young diagram. We now rewrite this formula by using the Frobenius coordinates and the min-max coordinates, and connect it with the transition measure. In order to regard the irreducible character values at a cycle as a function on $\mathbb{Y}$, set

$$\Sigma_k(\lambda) = \begin{cases} |\lambda|^{\downarrow k} \tilde{\chi}^\lambda_{(k,1^{|\lambda|-k})}, & |\lambda| \geqq k, \\ 0, & |\lambda| < k \end{cases} \tag{2.21}$$

for $k \in \mathbb{N}$ and $\lambda \in \mathbb{Y}$. In particular, $\Sigma_1(\lambda) = |\lambda|$.

Theorem 2.1 *For $k \in \mathbb{N}$ and $\lambda \in \mathbb{Y}$,*

$$\Sigma_k(\lambda) = -\frac{1}{k} [z^{-1}] \left\{ z^{\downarrow k} \frac{\Phi(z + \frac{1}{2}; \lambda)}{\Phi(z - k + \frac{1}{2}; \lambda)} \right\} \tag{2.22}$$

$$= -\frac{1}{k} [z^{-1}] \left\{ \frac{1}{G_{m_\lambda}(z) G_{m_\lambda}(z-1) \cdots G_{m_\lambda}(z-k+1)} \right\}. \tag{2.23}$$

Proof First we verify that the RHS of (2.22) is 0 if $|\lambda| < k$. In terms of the Frobenius coordinates $\lambda = (a_1, \ldots, a_d \mid b_1, \ldots, b_d)$,

$$z^{\downarrow k} \frac{\Phi(z + \frac{1}{2}; \lambda)}{\Phi(z - k + \frac{1}{2}; \lambda)} = z^{\downarrow k} \prod_{i=1}^{d} \frac{(z + \frac{1}{2} + b_i)(z - k + \frac{1}{2} - a_i)}{(z + \frac{1}{2} - a_i)(z - k + \frac{1}{2} + b_i)}. \tag{2.24}$$

The poles of (2.24) are all integers and satisfy

$$0 \leqq a_d - \frac{1}{2} < \cdots < a_1 - \frac{1}{2} < -b_1 + k - \frac{1}{2} < \cdots < -b_d + k - \frac{1}{2} \leqq k - 1$$

since $a_1 + b_1 \leqq |\lambda| < k$. Multiplied by $z^{\downarrow k}$, the denominator is then canceled. Hence (2.24) proves to be a polynomial in z.

Let us assume $|\lambda| \geqq k$. We show (2.22). By (1.8) and (2.21),

$$\Sigma_k(\lambda) = -\frac{1}{k} [z^{-1}] \left\{ z^{\downarrow k} \prod_{i=1}^{n} \frac{z - k - (\lambda_i + n - i)}{z - (\lambda_i + n - i)} \right\} \tag{2.25}$$

where $\lambda = (\lambda_i \geqq \lambda_2 \geqq \cdots)$ and $n = |\lambda|$. We note the equality

$$\Phi(z; \lambda) = \prod_{i=1}^{\infty} \frac{z - (-i + \frac{1}{2})}{z - (\lambda_i - i + \frac{1}{2})}. \tag{2.26}$$

In fact, multiplying both the numerator and the denominator in the RHS of (2.3) by $\prod_{c \in M(\lambda) \cap (-\mathbb{N}+\frac{1}{2})}(z - c)$, we get $\prod_{i=1}^{\infty} z - (-i + \frac{1}{2})$ as the new denominator. Now that $\lambda_{n+1} = 0$, (2.26) yields

$$\Phi(z - n + \frac{1}{2}; \lambda) = \prod_{i=1}^{n} \frac{z - n + i}{z - n - \lambda_i + i},$$

$$\Phi(z - n - k + \frac{1}{2}; \lambda) = \prod_{i=1}^{n} \frac{z - n - k + i}{z - n - k - \lambda_i + i},$$

and hence

$$\frac{\Phi(z - n + \frac{1}{2}; \lambda)}{\Phi(z - n - k + \frac{1}{2}; \lambda)} = \prod_{i=1}^{n} \frac{z - n + i}{z - n - k + i} \prod_{i=1}^{n} \frac{z - k - \lambda_i - n + i}{z - \lambda_i - n + i}.$$

Noting that the first product of the RHS is $z^{\downarrow k}/(z - n)^{\downarrow k}$, we have from (2.25)

$$\Sigma_k(\lambda) = -\frac{1}{k} [z^{-1}] \left\{ (z - n)^{\downarrow k} \frac{\Phi(z - n + \frac{1}{2}; \lambda)}{\Phi(z - n - k + \frac{1}{2}; \lambda)} \right\}. \tag{2.27}$$

For given λ and k, we can take a sufficiently large annulus $1 \ll |z| < \infty$ in which changing the contours $C \leftrightarrow C - n$ in the integral expressions is valid. Therefore, (2.22) follows from (2.27).

Finally, we verify the equality in (2.23). However, that is immediate from (2.7) and (2.16).

2.3 The Kerov–Olshanski Algebra

In this section, we focus on the algebra of polynomial functions in the coordinates of Young diagrams. Analysis of its structure in particular yields the Kerov polynomial and an asymptotic formula for irreducible characters of the symmetric groups.

We know two kinds of polynomials of 'degree' k as functions on $\mathbb{Y}$; one being $p_k(\lambda)$ of (2.2) in the Frobenius coordinates and the other $M_k(\tau_\lambda)$ of (2.10) in the min-max coordinates. Their generating functions of exponential type appear in (2.4) and (2.16)–(2.17) respectively. Since they are connected as (2.7), we can get the following relation between $\{p_k(\lambda)\}$ and $\{M_k(\tau_\lambda)\}$.

Proposition 2.4 *There exists an infinite matrix A satisfying*

- *A is upper-triangular*
- *All entries of A are nonnegative and rational*
- *All diagonal entries of A are equal to 1*

and

$$\left[M_2(\tau_\lambda) \; M_3(\tau_\lambda) \; M_4(\tau_\lambda) \; \cdots \right] = \left[2p_1(\lambda) \; 3p_2(\lambda) \; 4p_3(\lambda) \; \cdots \right] A. \qquad (2.28)$$

Proof We begin with (2.7) and use (2.3) and (2.6):

$$\frac{(1 - \frac{y_1}{z}) \cdots (1 - \frac{y_{r-1}}{z})}{(1 - \frac{x_1}{z}) \cdots (1 - \frac{x_r}{z})} = \prod_{i=1}^{d} \frac{(1 - \frac{-b_i}{z-(1/2)})(1 - \frac{a_i}{z+(1/2)})}{(1 - \frac{a_i}{z-(1/2)})(1 - \frac{-b_i}{z+(1/2)})} \qquad (2.29)$$

where $\lambda \in \mathbb{Y}$ has the Frobenius coordinates $(a_1, \ldots, a_d \,|\, b_1, \ldots, b_d)$ and the min-max coordinates $(x_1 < y_1 < \cdots < y_{r-1} < x_r)$. Expand logarithms of the both sides of (2.29) in $|z| \gg 1$. The LHS yields by (2.10) $\sum_{n=1}^{\infty} (M_n(\tau_\lambda)/n) z^{-n}$, while the RHS proceeds by (2.2) to

$$\sum_{k=1}^{\infty} \frac{p_k(\lambda)}{k} \left\{ \left(z - \frac{1}{2} \right)^{-k} - \left(z + \frac{1}{2} \right)^{-k} \right\} = \sum_{k=1}^{\infty} \frac{p_k(\lambda)}{k} z^{-k} \left\{ \left(1 - \frac{1}{2z} \right)^{-k} - \left(1 + \frac{1}{2z} \right)^{-k} \right\}$$

$$= \sum_{k=1}^{\infty} \frac{p_k(\lambda)}{k} z^{-k} \sum_{j=0}^{\infty} \binom{-k}{2j+1} \frac{-1}{2^{2j}} z^{-(2j+1)}$$

$$= \sum_{k=1}^{\infty} \frac{p_k(\lambda)}{k} z^{-k} \sum_{j=0}^{\infty} \frac{k^{\uparrow(2j+1)}}{(2j+1)! 2^{2j}} z^{-(2j+1)}$$

$$= \sum_{n=2}^{\infty} z^{-n} \sum_{0 \le j \le (n/2)-1} p_{n-2j-1}(\lambda) \frac{(n-1)^{\downarrow(2j)}}{(2j+1)! 2^{2j}}.$$

Hence we have

$$M_n(\tau_\lambda) = \sum_{0 \le j \le (n/2)-1} \binom{n}{2j+1} \frac{1}{2^{2j}} p_{n-2j-1}(\lambda), \qquad n \in \{2, 3, \ldots\},$$

which gives (2.28) and the other conditions for A.

Proposition 2.5 *Both* $\{p_n(\lambda)\}_{n\in\mathbb{N}}$ *and* $\{M_n(\tau_\lambda)\}_{n\in\{2,3,\cdots\}}$ *are algebraically independent.*

Proof [1]We show algebraic independence of $\{p_n(\lambda)\}_{n\in\mathbb{N}}$. Provided that

$$f(p_1(\lambda), \ldots, p_n(\lambda)) = \sum_{k_1, \ldots, k_n} \alpha_{k_1 \ldots k_n} p_1(\lambda)^{k_1} \ldots p_n(\lambda)^{k_n} = 0 \qquad (2.30)$$

holds for a polynomial f in n variables, let us show $f = 0$. In (2.30), the partial sum of the terms in which $k = k_1 + 2k_2 + \cdots + nk_n$ is maximal is denoted by $f^{\natural}$.

---

[1]The argument follows Proposition 1.5 in [16].

It suffices to verify that any coefficient $\alpha_{k_1 \cdots k_n}$ in $f^{\natural}$ vanishes because it then proves to be the case for all k's inductively. Let $x = (x_1, \ldots, x_l) \in \mathbb{R}^l$, $x_1 \geq \cdots \geq x_l > 0$, $l \geq k$, take $m \in \mathbb{N}$ and set $\lambda_i = \lfloor mx_i \rfloor$ for $i \in \{1, \ldots, l\}$, $\lambda = (\lambda_1 \geq \cdots \geq \lambda_l) \in \mathbb{Y}$. Putting this λ into (2.30), dividing the expression by the highest power of m and letting $m \to \infty$, we get

$$f^{\natural}(p_1(x), p_2(x), \ldots, p_n(x)) = 0 \tag{2.31}$$

where $p_j(x) = p_j(x_1, \ldots, x_l) = x_1^j + \cdots + x_l^j$ is the power sum in l variables. In fact, the effect of b_i's in the Frobenius coordinates and the other terms than $f^{\natural}$ tend to 0 as $m \to \infty$. Since $\{p_1(x_1, \ldots, x_l)^{k_1} \cdots p_n(x_1, \ldots, x_l)^{k_n} \mid k_1 + \ldots + nk_n = k \, (\leq l)\}$ is linearly independent by Proposition 1.4 (or a version of finite variables suffices), all coefficients in (2.31) is 0. This yields $f = 0$ and thus algebraic independence of $\{p_n(\lambda)\}_{n \in \mathbb{N}}$.

Provided that there exists an algebraic relation $g(M_2(\tau_\lambda), \ldots, M_{n+1}(\tau_\lambda)) = 0$ between $\{M_n(\tau_\lambda)\}_{n \in \{2,3,\ldots\}}$, rewrite it by using Proposition 2.4 as

$$g(2p_1(\lambda), \ldots, (n+1)p_n(\lambda)) + h(2p_1(\lambda), \ldots, (n+1)p_n(\lambda)) = 0.$$

By upper triangularity of A in (2.28), we get $g^{\natural}(2p_1(\lambda), \ldots, (n+1)p_n(\lambda)) = 0$ similarly to (2.31). Again through an inductive argument, we are led to $g = 0$. This completes the proof of algebraic independence of $\{M_n(\tau_\lambda)\}_{n \in \{2,3,\ldots\}}$.

The algebra $\mathbb{A}$ of functions on $\mathbb{Y}$ generated by $\{p_n(\lambda)\}_{n \in \mathbb{N}}$, or equivalently by $\{M_n(\tau_\lambda)\}_{n \in \{2,3,\ldots\}}$, is isomorphic to Λ of the symmetric functions. We call $\mathbb{A}$ the Kerov–Olshanski algebra after [20]. The two kinds of generators above induce the degrees of an element of $\mathbb{A}$. The canonical degree in $\mathbb{A}$ is defined by regarding $p_n(\lambda)$ as a homogeneous element of degree n. This is clearly the one inherited from Λ. On the other hand, the weight degree in $\mathbb{A}$ is defined by regarding $M_n(\tau_\lambda)$ as a homogeneous element of degree n. These degrees are denoted by deg and wt respectively: $\deg p_n(\lambda) = n$, $\text{wt}\, M_n(\tau_\lambda) = n$. If $f \in \mathbb{A}$ is not homogeneous, $\deg f$ and wt f indicate the degrees of the respective top homogeneous terms of f. For example, wt $p_n(\lambda) = n + 1$.

Recall that $\{M_n(\tau_\lambda)\}_{n \in \{2,3,\ldots\}}$ and $\{M_n(\mathfrak{m}_\lambda)\}_{n \in \{2,3,\ldots\}}$ are in polynomial relations to each other through (2.17) as was seen in Proposition 2.2. Actually, the relation is (a specialization of) the one between the power sums and the complete symmetric functions in Λ. Furthermore, moments of a probability on $\mathbb{R}$ are in polynomial relations to three kinds of cumulants, classical, free and Boolean, through the cumulant-moment formulas. In particular, we can take $\{M_n(\mathfrak{m}_\lambda)\}_{n \in \{2,3,\ldots\}}$ or $\{R_n(\mathfrak{m}_\lambda)\}_{n \in \{2,3,\ldots\}}$ as generators of $\mathbb{A}$. As is seen in the sequel, $\{\Sigma_k(\lambda)\}_{k \in \mathbb{N}}$ also generates $\mathbb{A}$. A key observation might be a resemblance between the two expressions (1.32) and (2.23). In the beginning, we have

$$\Sigma_1(\lambda) = R_2(\mathfrak{m}_\lambda) \, (= |\lambda|), \qquad \Sigma_2(\lambda) = R_3(\mathfrak{m}_\lambda). \tag{2.32}$$

Indeed, (2.11) and (2.19) yield

$$\Sigma_1(\lambda) = |\lambda| = \frac{1}{2} M_2(\tau_\lambda) = M_2(\mathfrak{m}_\lambda) = R_2(\mathfrak{m}_\lambda).$$

Moreover, (2.23) and (1.27) yield

$$\Sigma_2(\lambda) = -\frac{1}{2} [z^{-1}] \left\{ \left(z - \sum_{k=1}^{\infty} \frac{B_k(\mathfrak{m}_\lambda)}{z^{k-1}} \right) \left(z - 1 - \sum_{k=1}^{\infty} \frac{B_k(\mathfrak{m}_\lambda)}{(z-1)^{k-1}} \right) \right\}$$

$$= -\frac{1}{2} \left(-B_2(\mathfrak{m}_\lambda) - B_3(\mathfrak{m}_\lambda) + B_2(\mathfrak{m}_\lambda) - B_3(\mathfrak{m}_\lambda) \right)$$

$$= B_3(\mathfrak{m}_\lambda) = M_3(\mathfrak{m}_\lambda) = R_3(\mathfrak{m}_\lambda) \tag{2.33}$$

by noting $B_1(\mathfrak{m}_\lambda) = R_1(\mathfrak{m}_\lambda) = M_1(\mathfrak{m}_\lambda) = 0$.

Theorem 2.2 *For any $k \in \mathbb{N}$, $k \geq 3$, there exists a polynomial $P_k(x_2, \ldots, x_{k-1})$ in $k - 2$ variables satisfying*

$$\Sigma_k(\lambda) = R_{k+1}(\mathfrak{m}_\lambda) + P_k \big(R_2(\mathfrak{m}_\lambda), \ldots, R_{k-1}(\mathfrak{m}_\lambda) \big) \tag{2.34}$$

where a possible value of the weight degree of each term in the lower part

$$P_k \big(R_2(\mathfrak{m}_\lambda), \ldots, R_{k-1}(\mathfrak{m}_\lambda) \big)$$

belongs to $\{k - 1, k - 3, \ldots\}$ (every other integer) $\subset \mathbb{N}$.

Proof Let us write $G_\lambda = G_{\mathfrak{m}_\lambda}$, $M_k(\lambda) = M_k(\mathfrak{m}_\lambda)$, $R_k(\lambda) = R_k(\mathfrak{m}_\lambda)$, $B_k(\lambda) = B_k(\mathfrak{m}_\lambda)$ for short.

[*Step 1*] We will have an expression of $G_\lambda(z)^{-1} \ldots G_\lambda(z - k + 1)^{-1}$ in (2.23) in terms of the Laurent series in z in a similar way as (2.33). The expansion (1.27) of Boolean cumulant coefficients yields

$$\frac{1}{G_\lambda(z - r)} = z - r - \sum_{j=1}^{\infty} \frac{B_j(\lambda)}{(z-r)^{j-1}} = z - r - \sum_{j=1}^{\infty} \frac{B_j(\lambda)}{z^{j-1}} \left(\sum_{l=0}^{\infty} \frac{r^l}{z^l} \right)^{j-1} \tag{2.35}$$

for $r \in \{1, \ldots, k - 1\}$. Putting

$$\left(\sum_{l=0}^{\infty} t^l \right)^{j-1} = \sum_{l_1, \cdots, l_{j-1} = 0}^{\infty} t^{l_1 + \cdots + l_{j-1}} = \sum_{i=0}^{\infty} \alpha_{i, j-1} t^i,$$

$$\alpha_{i, j-1} = \big| \{ (l_1, \ldots, l_{j-1}) \in (\mathbb{N} \cup \{0\})^{j-1} \,|\, l_1 + \cdots + l_{j-1} = i \} \big|$$

into (2.35), we continue (2.35) as

$$= z - r - \sum_{j=1}^{\infty} \sum_{i=0}^{\infty} \frac{\alpha_{i,j-1} r^i B_j(\lambda)}{z^{i+j-1}} = z - r - \sum_{p=1}^{\infty} \frac{1}{z^{p-1}} \Big(\sum_{j=1}^{p} \alpha_{p-j,j-1} r^{p-j} B_j(\lambda) \Big)$$

$$= z - \sum_{p=1}^{\infty} A_{p,r}(\lambda) \frac{1}{z^{p-1}} \tag{2.36}$$

where

$$A_{p,r}(\lambda) = \begin{cases} \sum_{j=1}^{p} \alpha_{p-j,j-1} r^{p-j} B_j(\lambda), & p \geqq 2, \\ r + B_1(\lambda), & p = 1. \end{cases}$$

Since wt $B_j(\lambda) = $ wt $M_j(\lambda) = $ wt $M_j(\tau_\lambda) = j$ holds, we have wt $A_{p,r}(\lambda) = p$ and

$$A_{p,r}(\lambda) = B_p(\lambda) + (\text{wt-lower terms}), \qquad p \in \mathbb{N}. \tag{2.37}$$

[*Step 2*] Put (2.36) and (2.37) into $G_\lambda(z-r)^{-1}$ of (2.23):

$$\big(G_\lambda(z) G_\lambda(z-1) \cdots G_\lambda(z-k+1) \big)^{-1}$$

$$= \Big(z - \sum_{p=1}^{\infty} \frac{B_p(\lambda)}{z^{p-1}} \Big) \Big(z - \sum_{p=1}^{\infty} \frac{B_p(\lambda)}{z^{p-1}} + \sum_{p=1}^{\infty} \frac{*_1^p}{z^{p-1}} \Big) \cdots \Big(z - \sum_{p=1}^{\infty} \frac{B_p(\lambda)}{z^{p-1}} + \sum_{p=1}^{\infty} \frac{*_{k-1}^p}{z^{p-1}} \Big)$$

where $*_i^p, \cdots, *_{k-1}^p$ are terms of weight degree $\leqq p - 1$. Continue as

$$= \Big(z - \sum_{p=1}^{\infty} \frac{B_p(\lambda)}{z^{p-1}} \Big)^k + \sum_{j=1}^{k-1} \Big(z - \sum_{p=1}^{\infty} \frac{B_p(\lambda)}{z^{p-1}} \Big)^j \Big\{ \sum_{\sharp} \underbrace{\Big(\sum_{p=1}^{\infty} \frac{*}{z^{p-1}} \Big) \cdots \Big(\sum_{p=1}^{\infty} \frac{*}{z^{p-1}} \Big)}_{(k-j) \text{ product}} \Big\}$$

$$= G_\lambda(z)^{-k} + (\star) \tag{2.38}$$

where $(k-j)$ $*$'s are of weight degree $\leqq p-1$ though they are not identical. Moreover, $\sum_{\sharp}$ indicates a finite sum with the number depending k and j. Each j-term of $(\star)$ in (2.38) has such an expression as

$$z^i \frac{B_{p_1}(\lambda)}{z^{p_1-1}} \cdots \frac{B_{p_{j-i}}(\lambda)}{z^{p_{j-i}-1}} \frac{[\text{wt} \leqq q_1 - 1]}{z^{q_1-1}} \cdots \frac{[\text{wt} \leqq q_{k-j} - 1]}{z^{q_{k-j}-1}}, \qquad i \in \{0, 1, \ldots, j\}.$$

To pick up the term of z^{-1}, the requirement for the index is

$$i - \{ (p_1 - 1) + \cdots + (p_{j-i} - 1) + (q_1 - 1) + \cdots + (q_{k-j} - 1) \} = -1.$$

Then the weight degree of the coefficient is bounded by

$$p_1 + \cdots + p_{j-i} + (q_1 - 1) + \cdots + (q_{k-j} - 1) = j + 1 \leqq k.$$

We have thus $\mathrm{wt}\big([z^{-1}](\star)\big) \leqq k$. Combining this with (1.32), we get

$$\Sigma_k(\lambda) = -\frac{1}{k}[z^{-1}]\Big\{\frac{1}{G_\lambda(z)^k} + (\star)\Big\} = R_{k+1}(\lambda) + F(\lambda) \qquad (2.39)$$

with $F \in \mathbb{A}$, $\mathrm{wt}\, F \leqq k$.

[Step 3] Since $\{R_k(\lambda)\}$ generates $\mathbb{A}$, (2.39) yields existence of a polynomial P_k such that

$$\Sigma_k(\lambda) = R_{k+1}(\lambda) + P_k\big(R_2(\lambda), \ldots, R_k(\lambda)\big), \quad \mathrm{wt}\, P_k\big(R_2(\lambda), \ldots, R_k(\lambda)\big) \leqq k, \tag{2.40}$$

where it clearly suffices to take generators up to $R_k(\lambda)$ from the relations between generators of $\mathbb{A}$. Let us consider the involution

$$\mathrm{inv}(f)(\lambda) = f(\lambda'), \qquad f \in \mathbb{A}$$

induced by the transposition $\lambda \mapsto \lambda'$. Taking the character values at a k-cycle of $\lambda' \cong \lambda \otimes \mathrm{sgn}$, we have

$$\mathrm{inv}(\Sigma_k)(\lambda) = \Sigma_k(\lambda') = (-1)^{k-1}\Sigma_k(\lambda).$$

On the other hand, since the transition measure obeys $\mathfrak{m}_{\lambda'}(A) = \mathfrak{m}_\lambda(-A)$ for any Borel set A of $\mathbb{R}$, its moment satisfies $\mathrm{inv}(M_k)(\lambda) = M_k(\lambda') = (-1)^k M_k(\lambda)$. Then, (1.23) yields also for its free cumulant

$$\mathrm{inv}(R_k)(\lambda) = R_k(\lambda') = (-1)^k R_k(\lambda).$$

Taking inv of (2.40):

$$(-1)^{k-1}\Sigma_k(\lambda) = (-1)^{k+1}R_{k+1}(\lambda) + P_k\big(R_2(\lambda), \ldots, (-1)^k R_k(\lambda)\big)$$

and comparing it with (2.40), we have

$$P_k\big(R_2(\lambda), -R_3(\lambda), \ldots, (-1)^k R_k(\lambda)\big) = (-1)^{k-1} P_k\big(R_2(\lambda), R_3(\lambda), \ldots, R_k(\lambda)\big). \tag{2.41}$$

When k is even, (2.41) implies that the sum of the terms of even weight degree in $P_k\big(R_2(\lambda), \ldots, R_k(\lambda)\big)$ vanishes. Similarly, when k is odd, that of odd weight degree in $P_k\big(R_2(\lambda), \ldots, R_k(\lambda)\big)$ vanishes. Hence we conclude that possible weight degrees for the terms in $P_k\big(R_2(\lambda), \ldots, R_k(\lambda)\big)$ of (2.40) belong to $\{k-1, k-3, \ldots\}$. In particular, $R_k(\lambda)$ does not appear since there are no terms of weight degree k.

Seen from the viewpoint of the canonical degree, the following holds instead of Theorem 2.2.

Theorem 2.3 *For any $k \in \mathbb{N}$, there hold*

$$\Sigma_k(\lambda) = M_{k+1}(\mathfrak{m}_\lambda) + (\text{deg-}lower\ terms) \tag{2.42}$$
$$= p_k(\lambda) + (\text{deg-}lower\ terms). \tag{2.43}$$

Proof We first verify that the RHSs of (2.42) and (2.43) agree. Note $\deg M_k(\tau_\lambda) = k - 1$ by (2.28). The relation between $M_n(\mathfrak{m}_\lambda)$'s and $M_k(\tau_\lambda)$'s yield

$$M_{k+1}(\mathfrak{m}_\lambda) = \frac{1}{k+1} M_{k+1}(\tau_\lambda) + (\text{terms of deg} \leqq k - 1), \tag{2.44}$$

which together with (2.28) implies (2.42) agrees with (2.43). Needless to say, the terms of lower canonical degrees in both equations are not identical.

Next we show the equality of (2.42). In (2.34), the lower terms in the RHS satisfy wt $\leqq k - 1$ and deg $\leqq k - 2$. In the free cumulant-moment formula

$$R_{k+1}(\mathfrak{m}_\lambda) = \sum_{\pi \in \mathrm{NC}(k+1)} m_{\mathrm{NC}(k+1)}(\pi, 1_{k+1}) M_\pi(\mathfrak{m}_\lambda), \tag{2.45}$$

we have $\deg M_\pi(\mathfrak{m}_\lambda) = k + 1 - b(\pi)$ where $b(\pi)$ denotes the number of blocks of $\pi \in \mathrm{NC}(k+1)$. Indeed, (2.44) gives $\deg M_n(\mathfrak{m}_\lambda) = \deg M_n(\tau_\lambda) = n - 1$. Hence the term of the highest canonical degree in the RHS of (2.45) is the one of $b(\pi) = 1$, namely $M_{k+1}(\mathfrak{m}_\lambda)$. This completes the proof of (2.42). $\quad\blacksquare$

Corollary 2.1 *Both $\{\Sigma_k(\lambda)\}_{k \in \mathbb{N}}$ and $\{R_k(\mathfrak{m}_\lambda)\}_{k \in \{2,3,\dots\}}$ are algebraically independent.*

Corollary 2.2 *In Theorem 2.2, uniqueness of the polynomial P_k holds also. To be precise, the expression of (2.34) is unique without mentioning the weight degree of $P_k(R_2(\mathfrak{m}_\lambda), \dots, R_{k-1}(\mathfrak{m}_\lambda))$.*

Proof This follows from Corollary 2.1. $\quad\blacksquare$

Definition 2.1 Theorem 2.2, Corollary 2.2 and (2.32) determine the following sequence of polynomials:

$$K_2(x_2) = x_2, \quad K_3(x_2, x_3) = x_3,$$
$$K_{k+1}(x_2, \dots, x_{k+1}) = x_{k+1} + P_k(x_2, \dots, x_{k-1}), \quad k \geqq 3.$$

The polynomial K_k is called the Kerov polynomial.

Remark 2.1 The derivation of the Kerov polynomials based on comparing (1.32) and (2.23) is due to Okounkov as suggested in [3]. Carrying out Step 2 of the proof of Theorem 2.2, one has

$$K_4(x_2, x_3, x_4) = x_4 + x_2, \quad K_5(x_2, x_3, x_4, x_5) = x_5 + 5x_3, \quad \cdots.$$

The fact that all coefficients of the Kerov polynomials are positive integers is conjectured by Kerov and proved first by Féray [8]. Explicit forms of the first several Kerov polynomials are presented in [3].

Remark 2.2 Along the above discussion, Theorem 2.3 was proved by using the Kerov polynomials (Theorem 2.2), which does not seem to be optimal as readers might notice. It would be more natural to deduce (2.43) directly from (2.22) for a proof of Theorem 2.3.

Extending (2.21) to a general conjugacy class, set for $\rho \in \mathbb{Y}$

$$\Sigma_\rho(\lambda) = \begin{cases} |\lambda|^{\downarrow|\rho|}\, \tilde{\chi}^\lambda_{(\rho, 1^{|\lambda|-|\rho|})}, & |\lambda| \geqq |\rho|, \\ 0, & |\lambda| < |\rho|, \end{cases} \qquad \lambda \in \mathbb{Y}.$$

In particular, $\Sigma_\varnothing(\lambda) = 1$. As a linearizing formula, the following holds.

Proposition 2.6 *For* $\rho, \sigma \in \mathbb{Y}$,

$$\Sigma_\rho \Sigma_\sigma = \Sigma_{\rho \sqcup \sigma} + \sum_{\tau \in \mathbb{Y}: |\rho| \vee |\sigma| \leqq |\tau| < |\rho| + |\sigma|} a_\tau \Sigma_\tau, \qquad a_\tau \in \mathbb{Q}_{\geqq 0}.$$

Proposition 2.6 yields that $\Sigma_\rho \in \mathbb{A}$ and deg $\Sigma_\rho = |\rho|$. Hence we have

$$\Sigma_\rho = \Sigma_{\rho_1} \cdots \Sigma_{\rho_{l(\rho)}} + \text{(deg-lower terms)}. \tag{2.46}$$

Similarly for the weight degree also, the following holds.

Proposition 2.7 *For* $\rho, \sigma \in \mathbb{Y}$,

$$\Sigma_\rho \Sigma_\sigma = \Sigma_{\rho \sqcup \sigma} + \sum_{\tau \in \mathbb{Y}: |\tau| + l(\tau) \leqq |\rho| + l(\rho) + |\sigma| + l(\sigma) - 2} a_\tau \Sigma_\tau.$$

Hence we see wt $\Sigma_\rho = |\rho| + l(\rho)$ and

$$\Sigma_\rho = \Sigma_{\rho_1} \cdots \Sigma_{\rho_{l(\rho)}} + \text{(lower terms with weight degree} \leqq \text{wt } \Sigma_\rho - 2). \tag{2.47}$$

The expression (2.46) or (2.47) tells that $\{\Sigma_\rho\}_{\rho \in \mathbb{Y}}$ forms a basis of $\mathbb{A}$.

See [16] for the proofs of Proposition 2.6 and Proposition 2.7. In [13], we included their proofs based on partial permutations developed in [15].

Chapter 3
Continuous Diagram

Abstract In this chapter, continuous diagrams are introduced as limiting objects of the profiles of Young diagrams. It is important that the notion of a transition measure is extended for a continuous diagram.

3.1 Continuous Diagram I

Recall that $\mathbb{D}_0$ denotes the set of rectangular diagrams introduced in the beginning of Sect. 2.2. Extending the notion of a rectangular diagram, we consider a function $\omega : \mathbb{R} \longrightarrow \mathbb{R}$ satisfying:

(i) $|\omega(x_1) - \omega(x_2)| \leqq |x_1 - x_2|$, $\quad x_1, \, x_2 \in \mathbb{R}$

(ii) there exist $a < 0 < b$ such that $\omega(x) = |x|$ if $x \leqq a$ or $x \geqq b$,

and call such ω, or the graph $y = \omega(x)$, a (centered) continuous diagram. The set of continuous diagrams is denoted by $\mathbb{D}$. For $\omega \in \mathbb{D}$, the minimal closed interval $[a, b]$ satisfying (ii) is denoted by $\operatorname{supp} \omega$. It is obvious that $\mathbb{Y} \subset \mathbb{D}_0 \subset \mathbb{D}$ holds. Since the empty diagram $\varnothing \in \mathbb{Y}$ has the profile $y = |x|$, the definition implies $\operatorname{supp} \varnothing = \{0\}$.

The condition (i) for $\omega \in \mathbb{D}$ yields that ω is differentiable a.e. with $|\omega'(x)| \leqq 1$. Set

$$\mathbb{D}^{(a)} = \left\{ \omega \in \mathbb{D} \mid \operatorname{supp} \omega \subset (-a, a) \right\}, \qquad a > 0.$$

We obviously have $\mathbb{D} = \bigcup_{a>0} \mathbb{D}^{(a)} = \bigcup_{n\in\mathbb{N}} \mathbb{D}^{(n)}$. It is natural to equip $\mathbb{D}$ with the topology induced by the metric

$$\|\omega_1 - \omega_2\|_{\sup} = \sup_{x\in\mathbb{R}} |\omega_1(x) - \omega_2(x)|, \qquad \omega_1, \, \omega_2 \in \mathbb{D},$$

which is called the uniform (convergence) topology on $\mathbb{D}$. On the other hand, we can consider the inductive limit topology on $\mathbb{D} = \bigcup_{a>0} \mathbb{D}^{(a)}$ also where each $\mathbb{D}^{(a)}$ is given the relative topology of the uniform one on $\mathbb{D}$. Clearly, the inductive limit topology is stronger than the uniform topology on $\mathbb{D}$. Furthermore, the pointwise convergence topology on $\mathbb{D}$ is defined by the family of pseudo-metrics

$$\left\{ |\omega_1(x) - \omega_2(x)| \right\}_{x\in\mathbb{R}}, \qquad \omega_1, \, \omega_2 \in \mathbb{D}.$$

© The Author(s) 2016
A. Hora, *The Limit Shape Problem for Ensembles of Young Diagrams*,
SpringerBriefs in Mathematical Physics, DOI 10.1007/978-4-431-56487-4_3

Lemma 3.1 *The uniform and pointwise convergence topologies on $\mathbb{D}$ coincide.*

Proof Since the uniform topology is clearly stronger than the pointwise convergence one, let us show the converse implication. For any $\omega_0 \in \mathbb{D}$ given, take $a > 0$ such that $\omega_0 \in \mathbb{D}^{(a)}$. For any $\varepsilon > 0$, divide $[-a, a]$ to have $-a = x_0 < x_1 < \cdots < x_m = a$ with $\max_{i=1,\cdots,m}(x_i - x_{i-1}) \leqq \varepsilon/3$. If $\omega \in \mathbb{D}$ satisfies

$$\max_{i=1,\cdots,m}\left|\omega(x_i) - \omega_0(x_i)\right| \leqq \frac{\varepsilon}{3}, \tag{3.1}$$

then we have $\|\omega - \omega_0\|_{\sup} \leqq \varepsilon$ through the obvious triangular inequality. Indeed, since $\omega_0(x) = |x|$ holds for $x \in (-a, a)^c$, $|\omega(\pm a) - \omega_0(\pm a)| \leqq \varepsilon$ implies that $|\omega(x) - \omega_0(x)| \leqq \varepsilon$ holds for $x \in (-a, a)^c$ also. Hence the set of ω's satisfying (3.1) is included in the uniform ε-neighborhood of ω_0. This completes the proof.

3.2 Transition Measure II

We assigned to a rectangular diagram $\lambda \in \mathbb{D}_0$ its transition measure $\mathfrak{m}_\lambda$ by (2.15). It is characterized by the Stieltjes transform as in (2.16). There are several routes to reach the notion of the transition measure of a continuous diagram. Let us here review an elementary argument based on an approximation by rectangular diagrams.

For a given $\omega \in \mathbb{D}$, take $a > 0$ such that $\omega \in \mathbb{D}^{(a)}$. We can take a sequence $\{\lambda^{(n)}\}_{n\in\mathbb{N}} \subset \mathbb{D}_0 \cap \mathbb{D}^{(a)}$ converging to ω in $\mathbb{D}^{(a)}$. In fact, for any $\varepsilon > 0$, divide $[-a, a]$ as $-a = x_0 < x_1 < \cdots < x_m = a$ with $\max_{i=1,\cdots,m}(x_i - x_{i-1}) \leqq \varepsilon$. The oscillation of ω in $[x_{i-1}, x_i]$ is less than ε. Connect $(x_{i-1}, \omega(x_{i-1}))$ and $(x_i, \omega(x_i))$ by a portion of an element of $\mathbb{D}_0$ in such a way that the range lies between $\omega(x_{i-1})$ and $\omega(x_i)$. The resulting $\lambda \in \mathbb{D}_0 \cap \mathbb{D}^{(a)}$ then satisfies $\|\omega - \lambda\|_{\sup} \leqq \varepsilon$. Since $\operatorname{supp}\lambda^{(n)} \subset (-a, a)$, so are $\operatorname{supp}\mathfrak{m}_{\lambda^{(n)}}$ and $\operatorname{supp}\tau_{\lambda^{(n)}}$. Considering an approximation by polynomials on $[-a, a]$, we have for any $f \in C(\mathbb{R})$

$$\lim_{n\to\infty} \int_{\mathbb{R}} f(x)\left(\frac{\lambda^{(n)}(x) - |x|}{2}\right)' dx = \int_{\mathbb{R}} f(x)\left(\frac{\omega(x) - |x|}{2}\right)' dx. \tag{3.2}$$

Lemma 3.2 *For any $k \in \mathbb{N} \cup \{0\}$, $\left\{M_k(\mathfrak{m}_{\lambda^{(n)}})\right\}_{n\in\mathbb{N}}$ and $\left\{M_k(\tau_{\lambda^{(n)}})\right\}_{n\in\mathbb{N}}$ are both Cauchy sequences in $\mathbb{R}$.*

Proof It is trivial for $k = 0$. Since (2.10) yields

$$M_k(\tau_{\lambda^{(n)}}) = -\int_{\mathbb{R}} kx^{k-1}\left(\frac{\lambda^{(n)}(x) - |x|}{2}\right)' dx, \qquad k \in \mathbb{N},$$

$\left\{M_k(\tau_{\lambda^{(n)}})\right\}_{n\in\mathbb{N}}$ is of Cauchy as in (3.2). Since $M_k(\mathfrak{m}_{\lambda^{(n)}})$ is expressed by a polynomial in $M_j(\tau_{\lambda^{(n)}})$'s (not depending on n) by Proposition 2.2, $\left\{M_k(\mathfrak{m}_{\lambda^{(n)}})\right\}_{n\in\mathbb{N}}$ is of Cauchy also.

Combining Lemma 3.2 with a simple moment problem on a compact interval, we find a unique probability μ on $[-a, a]$ satisfying

$$M_k(\mu) = \lim_{n \to \infty} M_k(\mathfrak{m}_{\lambda^{(n)}}), \qquad k \in \mathbb{N} \cup \{0\}.$$

The probability μ does not depend on the choice of an approximating sequence $\{\lambda^{(n)}\} \subset \mathbb{D}_0 \cap \mathbb{D}^{(a)}$ of $\omega \in \mathbb{D}^{(a)}$. Indeed, for another approximating sequence $\{\mu^{(n)}\} \subset \mathbb{D}_0 \cap \mathbb{D}^{(a)}$, the limit of $M_k(\tau_{\mu^{(n)}})$ is determined by (3.2) and so is the one of $M_k(\mathfrak{m}_{\mu^{(n)}})$. Similarly, μ does not depend on the choice of $a > 0$ such that $\omega \in \mathbb{D}^{(a)}$ either. The probability μ thus determined for $\omega \in \mathbb{D}$ is called the (Kerov) transition measure of ω and denoted by $\mathfrak{m}_\omega$. Taking an approximating sequence $\{\lambda^{(n)}\}$ as above, we have

$$\lim_{n \to \infty} \int_{\mathbb{R}} f(x) \mathfrak{m}_{\lambda^{(n)}}(dx) = \int_{\mathbb{R}} f(x) \mathfrak{m}_\omega(dx), \qquad f \in C(\mathbb{R}) \tag{3.3}$$

through the approximation by polynomials on a compact interval.

Proposition 3.1 *The transition measure* $\mathfrak{m}_\omega$ *of* $\omega \in \mathbb{D}$ *satisfies*

$$\int_{\mathbb{R}} \frac{1}{z - x} \mathfrak{m}_\omega(dx) = \frac{1}{z} \exp\left\{ \int_{\mathbb{R}} \frac{1}{x - z} \left(\frac{\omega(x) - |x|}{2} \right)' dx \right\}, \qquad z \in \mathbb{C}^+. \tag{3.4}$$

Proof Take an approximating sequence $\{\lambda^{(n)}\} \subset \mathbb{D}_0 \cap \mathbb{D}^{(a)}$ of $\omega \in \mathbb{D}^{(a)}$. By (2.17), $\lambda^{(n)}$ satisfies the equality of (3.4) for $|z| > a$. Then, (3.2) and (3.3) yield the same equality for ω. Since the both sides are holomorphic in $\mathbb{C}^+$, we get (3.4). $\square$

For $a > 0$ and $\omega \in \mathbb{D}$, we set

$$\omega^a(x) = a^{-1} \omega(ax)$$

to have a rescaled diagram $\omega^a \in \mathbb{D}$.

Corollary 3.1 *For* $a > 0$ *and* $\omega \in \mathbb{D}$, *we have* $\mathfrak{m}_{\omega^a}(dx) = \mathfrak{m}_\omega(adx)$. *In particular,*

$$M_k(\mathfrak{m}_{\omega^a}) = a^{-k} M_k(\mathfrak{m}_\omega), \qquad k \in \mathbb{N} \cup \{0\}.$$

The above procedure of approximation does not work for defining an $\mathbb{R}$-valued measure though $\{M_k(\tau_{\lambda^{(n)}})\}_{n \in \mathbb{N}}$ is a Cauchy sequence. As in (2.9) for a rectangular diagram, we call

$$\tau_\omega = \left(\frac{\omega(x) - |x|}{2} \right)'' + \delta_0$$

the Rayleigh measure of $\omega \in \mathbb{D}$, provided that $(\omega(x) - |x|)'$ is of bounded variation.

Proposition 3.2 *If the Rayleigh measure τ_ω exists for $\omega \in \mathbb{D}$, it satisfies*

$$\omega(u) = \int_{\mathbb{R}} |u - x| \tau_\omega(dx), \qquad u \in \mathbb{R}, \qquad (3.5)$$

$$\frac{1}{G_{\mathfrak{m}_\omega}(z)} \frac{d}{dz} G_{\mathfrak{m}_\omega}(z) = -\int_{\mathbb{R}} \frac{1}{z - x} \tau_\omega(dx), \qquad z \in \mathbb{C}^+. \qquad (3.6)$$

Proof Note the bounded variation of $(\omega(x) - |x|)'$ and compactness of the supports of the measure considered. Then, a similar argument to Lemma 2.3 yields (3.5), and (3.6) easily follows from (3.4). ∎

Proposition 3.3 *Let probability μ on $\mathbb{R}$ with mean 0 and $\mathbb{R}$-valued measure τ on $\mathbb{R}$ have compact supports and satisfy*

$$\frac{1}{G_\mu(z)} \frac{d}{dz} G_\mu(z) = -\int_{\mathbb{R}} \frac{1}{z - x} \tau(dx), \qquad z \in \mathbb{C}^+.$$

Set

$$\omega(u) = \int_{\mathbb{R}} |u - x| \tau(dx), \qquad u \in \mathbb{R}. \qquad (3.7)$$

Then, $\omega \in \mathbb{D}$, $\mathfrak{m}_\omega = \mu$ and $\tau_\omega = \tau$ hold.

Proof We have

$$\frac{d}{dz}\left(\log G_\mu(z) + \int_{\mathbb{R}} \log(z - x) \tau(dx)\right) = 0$$

in $\mathbb{C}^+$. Here the argument of log is taken in $(-\pi, \pi)$. Since

$$\int_{\mathbb{R}} \log(z - x)\tau(dx) = \tau(\mathbb{R}) \log z + \int_{\mathbb{R}} \log\left(1 - \frac{x}{z}\right)\tau(dx),$$

we have

$$\log G_\mu(z) + \tau(\mathbb{R}) \log z + \int_{\mathbb{R}} \log\left(1 - \frac{x}{z}\right)\tau(dx) = c \ (= \text{const.}).$$

Letting $z \to \infty$, we see that $\tau(\mathbb{R}) = 1$ and $c = 0$. Hence

$$\int_{\mathbb{R}} \frac{1}{z - x} \mu(dx) = \frac{1}{z} \exp\left\{-\int_{\mathbb{R}} \log\left(1 - \frac{x}{z}\right)\tau(dx)\right\}, \qquad z \in \mathbb{C}^+ \qquad (3.8)$$

holds. Considering the Laurent expansions of (3.8) for $|z| \gg 1$ and comparing the coefficients of both sides, we get $M_1(\tau) = M_1(\mu) = 0$ in particular.

Take $a > 0$ such that supp $\tau \subset [-a, a]$. Setting $F(x) = \tau\big((-\infty, x]\big)$, we show

$$0 \leqq F(x) \leqq 1, \quad x \in \mathbb{R}. \tag{3.9}$$

For $z \in \mathbb{C}^+$, we have

$$\log G_\mu(z) = -\int_{[-a,a]} \log(z-x)\tau(dx) = -\int_{[-a,a]} \Big(\log z + \int_0^x \frac{-1}{z-t}dt\Big)\tau(dx)$$

$$= -\log z - \int_{-a}^0 \frac{F(t)}{z-t}dt + \int_0^a \frac{1-F(t)}{z-t}dt = -\log(z-a) - \int_{-a}^a \frac{F(t)}{z-t}dt. \tag{3.10}$$

Set $z = x + iy$ for $x \in (-a, a)$ and $y > 0$ in (3.10). We have

$$-\pi < \mathrm{Im}\, \log G_\mu(z) < 0, \quad \lim_{y\downarrow 0} \mathrm{Im}\, \log(z-a) = \pi,$$

and, if F is continuous at x,

$$F(x) = \lim_{y\downarrow 0}\Big(-\frac{1}{\pi}\mathrm{Im} \int_{-a}^a \frac{F(t)}{z-t}dt\Big).$$

Hence, letting $y \downarrow 0$ in (3.10), we get $-1 \leqq F(x) - 1 \leqq 0$. Noting that F is right continuous, has left limits and the continuous points of F are dense in $\mathbb{R}$, we have shown (3.9).

We show ω defined by (3.7) belongs to $\mathbb{D}$. Since $M_0(\tau) = 1$ and $M_1(\tau) = 0$ as verified above, it easily follows that $\omega(u) = |u|$ holds for $|u| \geqq a$. Computing ω' (as a Schwartz distribution) by using test functions, we get

$$\omega'(u) = \tau\big((-\infty, u]\big) - \tau\big((u, \infty)\big) = 2F(u) - 1. \tag{3.11}$$

Combining (3.11) with (3.9), we have $|\omega'(u)| \geq 1$ and hence $\omega \in \mathbb{D}$.

Differentiating (3.11) (as a Schwartz distribution) again, we see that ω has the Rayleigh measure and $\tau_\omega = \tau$ holds. Since τ_ω and $\mathfrak{m}_\omega$ are related in the same manner that τ and μ are in (3.8), $\mathfrak{m}_\omega$ and μ have the same Stieltjes transform. This implies μ is the transition measure $\mathfrak{m}_\omega$ of ω.

Remark 3.1 Changing the order of the integrals in (3.8), we have

$$\int_{\mathbb{R}} \log\Big(1 - \frac{x}{z}\Big)\tau(dx) = \int_{(-\infty,0)} \Big(\int_x^0 \frac{\frac{1}{z}}{1-\frac{y}{z}}dy\Big)\tau(dx) + \int_{(0,\infty)} \Big(\int_0^x \frac{-\frac{1}{z}}{1-\frac{y}{z}}dy\Big)\tau(dx)$$

$$= \int_{-\infty}^0 \frac{1}{z-y}F(y)dy - \int_0^\infty \frac{1}{z-y}\big(1 - F(y)\big)dy,$$

hence

$$\int_{\mathbb{R}} \frac{1}{z-x} \mu(dx) = \frac{1}{z} \exp\left(-\int_{-\infty}^{0} \frac{F(x)}{z-x} dx + \int_{0}^{\infty} \frac{1-F(x)}{z-x} dx\right), \quad z \in \mathbb{C}^{+}$$
(3.12)

with F satisfying $F(-\infty) = 0$, $F(\infty) = 1$ and $0 \leq F \leq 1$ as in (3.9). Similarly, (3.7) is rewritten as

$$\omega(u) = \int_{-\infty}^{u} F(x) dx + \int_{u}^{\infty} \left(1 - F(x)\right) dx, \quad u \in \mathbb{R}.$$
(3.13)

Since (3.11) holds in this case also, we have $\omega \in \mathbb{D}$. Actually, it can be shown that, if $\mu \in \mathscr{P}(\mathbb{R})$ has mean 0 and compact support, then there exists a unique $\omega \in \mathbb{D}$ such that $\mu = \mathfrak{m}_{\omega}$ which is characterized by (3.12) and (3.13), though τ_{ω} may not necessarily exist. Including the case where μ has non-compact support (with a certain moment condition), [18] gives a thorough treatment of interplay between μ, F and ω.

In the approach to the limit shape problem that we will adopt, limiting objects are often captured first in terms of free cumulants, then the R-transform, the Stieltjes transform of the desired probability, and finally the corresponding continuous diagram. Proposition 3.3 is useful in such a context. We now give two fundamental examples—free counterparts of Gauss and Poisson distributions.

Example 3.1 Wigner's semi-circle distribution with mean m and variance v:

$$\frac{1}{2\pi v} \sqrt{4v - (x - m)^2} \, 1_{[m-2\sqrt{v}, m+2\sqrt{v}]}(x) dx$$
(3.14)

is characterized by the free cumulant sequence: $R_1 = m$, $R_2 = v$, $R_3 = R_4 = \cdots = 0$. Since the transition measure of a continuous diagram has mean 0, let us start from the free cumulant sequence

$$R_1 = 0, \quad R_2 = v, \quad R_3 = R_4 = \cdots = 0, \quad v > 0$$
(3.15)

and compute the corresponding continuous diagram. It immediately follows from (3.15) and 1.26 that $z = K(\zeta) = \zeta^{-1} + v\zeta$, which is inverted as

$$\zeta = G(z) = \frac{z - \sqrt{z^2 - 4v}}{2v}, \quad z \in \mathbb{C}^{+}$$
(3.16)

by noting that $\lim_{z \to \infty} zG(z) = 1$ holds. For $\sqrt{A}$ in (3.16), the argument of A is taken in $(0, 2\pi)$. Hence we have

$$\frac{G'(z)}{G(z)} = -\frac{1}{\sqrt{z^2 - 4v}} = -\int_{-2\sqrt{v}}^{2\sqrt{v}} \frac{dx}{\pi \sqrt{4v - x^2}}.$$
(3.17)

The second equality in (3.17) is well known, or obtained by using the Stieltjes inversion formula. Applying Proposition 3.3, we get

$$\omega(u) = \int_{-2\sqrt{v}}^{2\sqrt{v}} |u - x| \frac{dx}{\pi\sqrt{4v - x^2}} = \begin{cases} \frac{2}{\pi}\left(u\arcsin\frac{u}{2\sqrt{v}} + \sqrt{4v - u^2}\right), & |u| \leq 2\sqrt{v}, \\ |u|, & |u| > 2\sqrt{v}. \end{cases}$$
(3.18)

The particular case of $m = 0$ and $v = 1$ in (3.14) gives the standard semi-circle distribution. The corresponding continuous diagram of (3.18) is what we call the limit shape (see Fig. 4.1).

Example 3.2 The free Poisson distribution, or the Marchenko–Pastur distribution, with parameter λ:

$$\begin{cases} v, & \lambda \geq 1, \\ v + (1 - \lambda)\delta_0, & 0 < \lambda \leq 1, \end{cases}$$

$$v(dx) = \frac{\sqrt{4\lambda - (x - 1 - \lambda)^2}}{2\pi x} 1_{\left((1-\sqrt{\lambda})^2, (1+\sqrt{\lambda})^2\right)}(x)dx$$

is characterized by the free cumulant sequence: $R_1 = R_2 = \cdots = \lambda$. Translating it to have mean 0, we start from

$$R_1 = 0, \quad R_2 = R_3 = \cdots = \lambda > 0 \tag{3.19}$$

to compute the corresponding continuous diagram. Similarly to (3.16) we have

$$z = K(\zeta) = \frac{1}{\zeta} + \frac{\lambda\zeta}{1 - \zeta}, \qquad \zeta = G(z) = \frac{z + 1 - \sqrt{(z - 1)^2 - 4\lambda}}{2(z + \lambda)}$$

from (3.19) and (1.26), and hence

$$-\frac{G'(z)}{G(z)} = \frac{1}{2\sqrt{(z - 1)^2 - 4\lambda}}\left(1 + \frac{\lambda - 1}{z + \lambda}\right) + \frac{1}{2(z + \lambda)}. \tag{3.20}$$

We seek $\mathbb{R}$-valued measure τ on $\mathbb{R}$ with compact support which has (3.20) as its Stieltjes transform. Set $z = x + iy$ ($y > 0$) and $\sqrt{(z - 1)^2 - 4\lambda} = u + iv$. Then, the imaginary part of (3.20) is given by

$$-\frac{1}{2}\left\{\frac{v}{u^2 + v^2} + \frac{(\lambda - 1)(uy + v(x + \lambda))}{(u(x + \lambda) - vy)^2 + (uy + v(x + \lambda))^2} + \frac{y}{(x + \lambda)^2 + y^2}\right\}. \tag{3.21}$$

Since $(z - 1)^2 - 4\lambda = (x - 1)^2 - 4\lambda - y^2 + i2y(x - 1)$, (3.21) tends to 0 as $y \downarrow 0$ if $(x - 1)^2 - 4\lambda > 0$ and $x + \lambda \neq 0$. On the other hand, if $(x - 1)^2 - 4\lambda < 0$, we get

$$\lim_{y\downarrow 0}(3.21) = -\frac{1}{2}\Big(\frac{1}{\sqrt{4\lambda - (x-1)^2}} + \frac{\lambda - 1}{(x+\lambda)\sqrt{4\lambda - (x-1)^2}}\Big)$$

by noting $\lim_{y\downarrow 0} u = 0$. The Stieltjes inversion formula then yields that τ (if it exists) has the absolutely continuous part

$$\tilde{\tau}(dx) = \frac{1}{2\pi}\Big(\frac{1}{\sqrt{4\lambda - (x-1)^2}} + \frac{\lambda - 1}{(x+\lambda)\sqrt{4\lambda - (x-1)^2}}\Big)1_{(1-2\sqrt{\lambda},1+2\sqrt{\lambda})}(x)dx.$$
(3.22)

The Stieltjes transform of (3.22) is actually computed (by any method you like, e.g. residue calculus) as

$$G_{\tilde{\tau}}(z) = (1 - \delta_{\lambda 1})\frac{\lambda - 1}{2}\Big\{\frac{1}{|\lambda - 1|(z+\lambda)} + \frac{1}{(z+\lambda)\sqrt{(z-1)^2 - 4\lambda}}\Big\}$$
$$+ \frac{1}{2\sqrt{(z-1)^2 - 4\lambda}}, \qquad z \in \mathbb{C}^+. \quad (3.23)$$

Comparing (3.23) with (3.20), we see that the $\mathbb{R}$-valued measure

$$\tau = \begin{cases} \tilde{\tau}, & \lambda > 1, \\ \tilde{\tau} + \frac{1}{2}\delta_{-1}, & \lambda = 1, \\ \tilde{\tau} + \delta_{-\lambda}, & 0 < \lambda < 1 \end{cases} \qquad (3.24)$$

defined from $\tilde{\tau}$ of (3.22) is compactly supported and has (3.20) as its Stieltjes transform. Applying Proposition 3.3, we get the corresponding continuous diagram from the Rayleigh measure τ of (3.24). As for $\tilde{\tau}$,

$$\int_{\mathbb{R}} |u - x|\tilde{\tau}(dx)$$
$$= \frac{\sqrt{\lambda}}{2\pi}\int_{-2}^{2}\Big|\frac{u-1}{\sqrt{\lambda}} - x\Big|\frac{dx}{\sqrt{4 - x^2}} + \frac{\lambda - 1}{2\pi}\int_{-2}^{2}\Big|\frac{u-1}{\sqrt{\lambda}} - x\Big|\frac{dx}{(x + \sqrt{\lambda} + \frac{1}{\sqrt{\lambda}})\sqrt{4 - x^2}}.$$

The case of $\lambda = 1$ is easier. To handle the case of $\lambda \neq 1$, use (if you like)

$$\frac{1}{(x + \sqrt{\lambda} + \frac{1}{\sqrt{\lambda}})\sqrt{4 - x^2}} = \frac{d}{dx}\Big(\frac{2}{|\sqrt{\lambda} - \frac{1}{\sqrt{\lambda}}|}\arcsin\frac{\sqrt{(\sqrt{\lambda} + \frac{1}{\sqrt{\lambda}} + 2)(x+2)}}{2\sqrt{x + \sqrt{\lambda} + \frac{1}{\sqrt{\lambda}}}}\Big),$$

$$\frac{x}{(x + \sqrt{\lambda} + \frac{1}{\sqrt{\lambda}})\sqrt{4 - x^2}} = \frac{1}{\sqrt{4 - x^2}} - \frac{\sqrt{\lambda} + \frac{1}{\sqrt{\lambda}}}{(x + \sqrt{\lambda} + \frac{1}{\sqrt{\lambda}})\sqrt{4 - x^2}}.$$

The result of $\omega(u) = \int_{\mathbb{R}} |u - x| \tau(dx)$ is as follows: if $\lambda \geq 1$,

$$\omega(u) = \begin{cases} \dfrac{\sqrt{4\lambda - (u-1)^2}}{\pi} + \dfrac{u-\lambda}{\pi} \arcsin \dfrac{u-1}{2\sqrt{\lambda}} - \dfrac{u+\lambda}{2} \\ \quad + \dfrac{2(u+\lambda)}{\pi} \arcsin \dfrac{\sqrt{(\sqrt{\lambda}+\frac{1}{\sqrt{\lambda}}+2)(u-1+2\sqrt{\lambda})}}{2\sqrt{u+\lambda}}, \quad 1 - 2\sqrt{\lambda} \leq u \leq 1 + 2\sqrt{\lambda}, \\ |u|, \quad\quad\quad\quad\quad\quad\quad\quad u \leq 1 - 2\sqrt{\lambda} \text{ or } 1 + 2\sqrt{\lambda} \leq u, \end{cases}$$

and, if $0 < \lambda \leq 1$,

$$\omega(u) = \begin{cases} -u, & u \leq -\lambda, \\ u + 2\lambda, & -\lambda \leq u \leq 1 - 2\sqrt{\lambda}, \\ \dfrac{\sqrt{4\lambda - (u-1)^2}}{\pi} + \dfrac{u-\lambda}{\pi} \arcsin \dfrac{u-1}{2\sqrt{\lambda}} + \dfrac{3(u+\lambda)}{2} \\ \quad - \dfrac{2(u+\lambda)}{\pi} \arcsin \dfrac{\sqrt{(\sqrt{\lambda}+\frac{1}{\sqrt{\lambda}}+2)(u-1+2\sqrt{\lambda})}}{2\sqrt{u+\lambda}}, & 1 - 2\sqrt{\lambda} \leq u \leq 1 + 2\sqrt{\lambda}, \\ u, & 1 + 2\sqrt{\lambda} \leq u. \end{cases}$$

3.3 Continuous Diagram II

This section contains remarks on the topologies on $\mathbb{D}$.

Lemma 3.3 *The following three families of pseudo-distances on $\mathbb{D}$ give the same topology: for $\omega_1, \omega_2 \in \mathbb{D}$,*

$$\left\{ |M_k(\mathfrak{m}_{\omega_1}) - M_k(\mathfrak{m}_{\omega_2})| \right\}_{k \in \mathbb{N}},$$

$$\left\{ \left| \int_{\mathbb{R}} x^k (\omega_1(x) - |x|)' dx - \int_{\mathbb{R}} x^k (\omega_2(x) - |x|)' dx \right| \right\}_{k \in \mathbb{N}},$$

$$\left\{ \left| \int_{\mathbb{R}} x^{k-1} (\omega_1(x) - \omega_2(x)) dx \right| \right\}_{k \in \mathbb{N}}.$$

Proof The formula (3.4) is rewritten for $|z| \gg 1$ as

$$\sum_{n=0}^{\infty} \frac{M_n(\mathfrak{m}_{\omega})}{z^n} = \exp \left\{ -\sum_{k=1}^{\infty} \frac{1}{z^{k+1}} \int_{\mathbb{R}} x^k \left(\frac{\omega(x) - |x|}{2} \right)' dx \right\}. \tag{3.25}$$

Hence $\{M_n(\mathfrak{m}_{\omega})\}_n$ and $\{\int_{\mathbb{R}} x^k (\omega(x) - |x|)' dx\}_k$ are expressed by polynomials in each other. The rest follows immediately from integration by parts.

The beginning of the relation given by (3.25) are

$$M_0(\mathfrak{m}_{\omega}) = 1, \quad M_1(\mathfrak{m}_{\omega}) = 0, \quad M_2(\mathfrak{m}_{\omega}) = \int_{\mathbb{R}} \frac{\omega(x) - |x|}{2} dx, \quad \cdots. \tag{3.26}$$

We call the topology determined in Lemma 3.3 the moment topology on $\mathbb{D}$.

Lemma 3.4 *For $a > 0$, the moment and uniform topologies on $\mathbb{D}^{(a)}$ coincide.*

Proof We verify that the moment topology is stronger since the converse is immediate from a triangle inequality. By virtue of Lemma 3.1, the uniformity can be replaced by the pointwise convergence. For $\omega_1, \omega_2 \in \mathbb{D}$, we have

$$|\omega_1(u) - \omega_2(u)| = \left| \int_{-a}^{a} 1_{[-a,u]}(x)\{(\omega_1(x) - |x|)' - (\omega_2(x) - |x|)'\}dx \right|, \quad u \in [-a, a].$$

Approximating $1_{[-a,u]}$ by a continuous function and then by a polynomial, we see that, for any $\delta > 0$, there exists a polynomial p_δ such that

$$|\omega_1(u) - \omega_2(u)| \leqq \delta + \left| \int_{-a}^{a} p_\delta(x)\{(\omega_1(x) - |x|)' - (\omega_2(x) - |x|)'\}dx \right|.$$

The second term of RHS can be arbitrarily small in the moment topology.

Proposition 3.4 *The moment topology is stronger than the uniform topology on $\mathbb{D}$.*

Proof Since both topologies are metrizable, it suffices to discuss convergence of a sequence. Assume that $\{\omega_n\}_{n=1}^{\infty} \subset \mathbb{D}$ converges to $\omega_0 \in \mathbb{D}$ in the moment topology. First let $\omega_0(x) \equiv |x|$. If ω_n does not converge uniformly, then $\exists \varepsilon > 0, \exists\{\omega_{n_j}\}_j$: subsequence, $\forall j, \sup_{x\in\mathbb{R}}|\omega_{n_j}(x) - |x|| \geqq \varepsilon$. Since $\omega_{n_j} \in \mathbb{D}$, this condition necessarily yields $\omega_{n_j}(0) \geqq \varepsilon$. We can take a triangular diagram[1] Δ satisfying $\omega_{n_j}(x) \geqq \Delta(x)$ ($\forall x \in \mathbb{R}$) and have

$$M_2(\mathfrak{m}_{\omega_{n_j}}) = \int_{\mathbb{R}} \frac{\omega_{n_j}(x) - |x|}{2}dx \geqq \int_{\mathbb{R}} \frac{\Delta(x) - |x|}{2}dx > 0 = M_2(\mathfrak{m}_{\omega_0}).$$

This contradicts the convergence of the moments.

Let $\omega_0(x) \not\equiv |x|$ hence $\int_{\mathbb{R}}(\omega_0(x) - |x|)dx > 0$ hold. By the assumption,

$$\lim_{n\to\infty} \int_{\mathbb{R}} x^{k-1}(-x)(\omega_n(x) - |x|)'dx = \int_{\mathbb{R}} x^{k-1}(-x)(\omega_0(x) - |x|)'dx, \quad k \in \mathbb{N}. \tag{3.27}$$

Setting

$$c_n = \int_{\mathbb{R}} (-x)(\omega_n(x) - |x|)'dx, \quad n \in \mathbb{N} \cup \{0\},$$

we can assume $c_n > 0$ since $\lim_{n\to\infty} c_n = c_0 > 0$. Set

$$\nu_n(dx) = c_n^{-1}(-x)(\omega_n(x) - |x|)'dx, \quad n \in \mathbb{N} \cup \{0\}. \tag{3.28}$$

---

[1]A continuous diagram Δ such that the region $|x| \leqq y \leqq \Delta(x)$ is a triangle.

Since $\omega_n \in \mathbb{D}$, we have $\nu_n \in \mathscr{P}(\mathbb{R})$ and convergence of the moments of ν_n by (3.27), which implies the weak convergence of ν_n to ν_0 as $n \to \infty$ (because ν_0 is compactly supported). Noting that $\nu_0((0, \infty)) > 0$ and $\nu_0((-\infty, 0)) > 0$ hold, we first consider on $(0, \infty)$. In (3.28), we want to remove the effect of multiplication by x. This turns out to be possible since we have $|(\omega_n(x) - |x|)'| \leq 2$ and hence

$$0 \leqq \int_0^\delta \{-(\omega_n(x) - |x|)'\}dx \leqq 2\delta$$

for any $\delta > 0$ and $n \in \mathbb{N} \cup \{0\}$. Therefore,

$$\lim_{n \to \infty} \int_0^\infty f(x)\{-(\omega_n(x) - |x|)'\}dx = \int_0^\infty f(x)\{-(\omega_0(x) - |x|)'\}dx, \quad f \in C_b(\mathbb{R}).$$

A similar discussion proceeds on $(-\infty, 0)$ also. Then, we get pointwise convergence of the distribution functions and consequently

$$\lim_{n \to \infty} \omega_n(u) = \omega(u), \quad u \in \mathbb{R}.$$

This with Lemma 3.1 completes the proof.

Remark 3.2 We thus have three topologies on $\mathbb{D} = \bigcup_{a>0} \mathbb{D}^{(a)}$ according to the order of strength: the inductive limit topology (of the uniform topology on each), the moment topology, and the uniform topology.

Chapter 4
Static Model

Abstract Since the Plancherel measure on the path space of the Young graph is ergodic, some deterministic aspects will appear in a macroscopic point of view. In this chapter, we describe the famous limit shape problem for the profiles of random Young diagrams in the Plancherel ensemble, which is due to Vershik–Kerov and Logan–Shepp, as a strong law of large numbers for the Plancherel measure on the path space. Free cumulants of the transition measure play a central role in characterizing the limit shape of Young diagrams.

4.1 Balanced Young Diagrams

The Plancherel measure M_{Pl} was introduced by (1.17) as an ergodic central probability on the path space $\mathfrak{T}$. The aim of this section is to show the following asymptotic property of M_{Pl}. Recall $t(n) \in \mathbb{Y}_n$ denotes the nth vertex of the path $t \in \mathfrak{T}$.

Theorem 4.1 *There exists $c > 0$ such that, for M_{Pl}-a.s. path $t \in \mathfrak{T}$,*

$$t(n)_1 \leqq c\sqrt{n} \quad and \quad t(n)_1' \leqq c\sqrt{n}$$

hold if n is sufficiently large.

Let us recall the Robinson–Schensted correspondence holding between permutations and pairs of standard tableaux. To obtain from $x = \begin{pmatrix} 1 & 2 & \ldots & n \\ x_1 & x_2 & \ldots & x_n \end{pmatrix} \in \mathfrak{S}_n$ a pair (P, Q) of standard tableaux with a common shape in $\mathbb{Y}_n$, we consider a sequence of pairs

$$(P_0, Q_0) = (\varnothing, \varnothing), (P_1, Q_1), \ldots, (P_n, Q_n) = (P, Q)$$

as follows.

© The Author(s) 2016
A. Hora, *The Limit Shape Problem for Ensembles of Young Diagrams*,
SpringerBriefs in Mathematical Physics, DOI 10.1007/978-4-431-56487-4_4

1. P_1 contains x_1 in the box $\square \in \mathbb{Y}_1$.
2. If $x_2 > x_1$, P_2 has shape (2^1) with x_1, x_2 in the row. If $x_2 < x_1$, x_2 bumps x_1 and send it to the next row to obtain P_2 of shape (1^2) with x_2, x_1 in the column.
3. Provided that P_{k-1} containing $x_1, \ldots, x_{k-1}$ is in hand, compare x_k with the letters in the first row R_1 of P_{k-1}. If x_k is larger than any other, simply add x_k at the right end of R_1 to get P_k. Otherwise, x_k bumps the smallest y larger than x_k in R_1 and send y to the next row. Then, compare y with the second row of P_{k-1}, and continue such a bumping procedure until x_n is put to get P_n finally.
4. Letting $\lambda^{(k)} \in \mathbb{Y}_k$ denote the shape of P_k, we have a path $\lambda^{(0)} = \varnothing \nearrow \lambda^{(1)} \nearrow \ldots \nearrow \lambda^{(k)}$. Let $Q_k \in \mathrm{STab}(\lambda^{(k)})$ correspond to this path. We thus get $P = P_n$, $Q = Q_n \in \mathrm{STab}(\lambda^{(n)})$ with $\lambda^{(n)} \in \mathbb{Y}_n$.

Proposition 4.1 *For $n \in \mathbb{N}$, the map $x \to (P, Q)$ gives a bijection between*

$$\mathfrak{S}_n \cong \left\{ (P, Q) \,\middle|\, P, Q \in \mathrm{STab}(\lambda), \ \lambda \in \mathbb{Y}_n \right\}. \tag{4.1}$$

Moreover, letting $L_n(x)$ be the length of a longest increasing subsequence of x, we have $L_n(x) = \lambda_1$ under (4.1).

See [25, Chap. 3] for the proof of Proposition 4.1. Here $(x_{i_1} x_{i_2} \ldots x_{i_k})$ is called an increasing subsequence of $x = \begin{pmatrix} 1 & 2 & \ldots & n \\ x_1 & x_2 & \ldots & x_n \end{pmatrix} \in \mathfrak{S}_n$ if $x_{i_1} < x_{i_2} < \cdots < x_{i_k}$ holds with $i_1 < i_2 < \cdots < i_k$.

Lemma 4.1 *The law of L_n with respect to the uniform probability Prob_n on $\mathfrak{S}_n$ is given by*

$$\mathrm{Prob}_n(L_n = l) = M_{\mathrm{Pl}}^{(n)}\big(\{\lambda \in \mathbb{Y}_n \,|\, \lambda_1 = l\}\big), \qquad l \in \{1, 2, \ldots, n\}. \tag{4.2}$$

Proof Propositions 4.1 and 1.2 yield

$$\big|\{x \in \mathfrak{S}_n \,|\, L_n(x) = l\}\big| = \big|\{(P, Q) \,|\, P, Q \in \mathrm{STab}(\lambda), \ \lambda \in \mathbb{Y}_n, \ \lambda_1 = l\}\big|$$
$$= \sum_{\lambda \in \mathbb{Y}_n : \lambda_1 = l} (\dim \lambda)^2.$$

Multiplying this by $1/n!$, we get (4.2).

Lemma 4.2 *For $n \in \mathbb{N}$,*

$$\mathrm{Prob}_n(L_n \geqq l) \leqq n^{\downarrow l}/(l!)^2, \qquad l \in \{1, 2, \ldots, n\}. \tag{4.3}$$

Proof Given $l \in \{1, 2, \cdots, n\}$, estimate the number of permutations x with $L_n(x) \geqq l$ quite roughly. Since there may be $\binom{n}{l}$ choices for l positions and l letters to extract an increasing subsequence of length l, we have

$$\mathrm{Prob}_n(L_n \geqq l) \leqq \frac{1}{n!} \binom{n}{l}^2 (n - l)! = \frac{n^{\downarrow l}}{(l!)^2}.$$

For $c > 0$, we consider

$$\sum_{n=1}^{\infty} M_{\mathrm{Pl}}\big(\{t \in \mathfrak{T} \mid t(n)_1 > c\sqrt{n}\}\big) = \sum_{n=1}^{\infty} M_{\mathrm{Pl}}^{(n)}\big(\{\lambda \in \mathbb{Y}_n \mid \lambda_1 > c\sqrt{n}\}\big). \qquad (4.4)$$

Combining (4.2) and (4.3) with Stirling's formula, we easily see (4.4) is finite if $c > e$. Since the same estimate holds for $t(n)_1'$ by the symmetry of the Plancherel measure, we have

$$\sum_{n=1}^{\infty} M_{\mathrm{Pl}}\big(\{t \in \mathfrak{T} \mid t(n)_1 > c\sqrt{n} \ \text{or} \ t(n)_1' > c\sqrt{n}\}\big) < \infty$$

if $c > e$. Then, the Borel–Cantelli lemma completes the proof of Theorem 4.1.

Since any $\lambda \in \mathbb{Y}_n$ satisfies $\lambda_1 \lambda_1' \geq n$, the condition $\lambda_1 \leq c\sqrt{n}$, $\lambda_1' \leq c\sqrt{n}$ for $\lambda \in \mathbb{Y}_n$ implies

$$\frac{\sqrt{n}}{c} \leq \lambda_1 \leq c\sqrt{n}, \quad \frac{\sqrt{n}}{c} \leq \lambda_1' \leq c\sqrt{n}. \qquad (4.5)$$

Such a $\lambda \in \mathbb{Y}_n$ satisfying (4.5) is said to be c-balanced.

4.2 Convergence to the Limit Shape

In this section, we show a fundamental fact about the asymptotic property of the profiles of Young diagrams in the Plancherel ensemble (Theorem 4.2). Let Ω denote the continuous diagram of (3.18) in the case of $\nu = 1$ (Fig. 4.1):

$$\Omega(x) = \begin{cases} \frac{2}{\pi}(x\arcsin\frac{x}{2} + \sqrt{4 - x^2}), & |x| \leq 2, \\ |x|, & |x| > 2. \end{cases} \qquad (4.6)$$

Since Theorem 4.1 assures that typical Young diagrams grow in a balanced manner (4.5) under the Plancherel measure, it is appropriate to consider a limit rescaled by $1/\sqrt{n}$. For $\lambda \in \mathbb{Y}_n \subset \mathbb{D}_0$, define $\lambda^{\sqrt{n}} \in \mathbb{D}_0$ by

$$\lambda^{\sqrt{n}}(x) = \frac{1}{\sqrt{n}}\lambda(\sqrt{n}x). \qquad (4.7)$$

Theorem 4.2 *For M_{Pl}-a.s. path $t \in \mathfrak{T}$, rectangular diagram $t(n)^{\sqrt{n}}$ converges to Ω as $n \to \infty$ in $\mathbb{D}$ with respect to the moment topology, hence to the uniform topology also.*

A weak law obviously follows from Theorem 4.2, a strong law. Recall that $M_{\mathrm{Pl}}^{(n)}$ denotes the Plancherel measure (1.18) on $\mathbb{Y}_n$ as a marginal distribution of M_{Pl}.

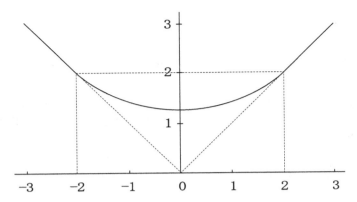

Fig. 4.1 Limit shape Ω of Young diagram

Corollary 4.1 *For any $\varepsilon > 0$,*

$$\lim_{n\to\infty} M_{\mathrm{Pl}}^{(n)}\left(\left\{\lambda \in \mathbb{Y}_n \,\middle|\, \sup_{x\in\mathbb{R}} |\lambda^{\sqrt{n}}(x) - \Omega(x)| \geq \varepsilon\right\}\right) = 0.$$

Since convergence of the moments is equivalent to convergence of the free cumulants by the free cumulant-moment formula (1.23), Theorem 4.2 follows from the following result.

Theorem 4.3 *For M_{Pl}-a.s. path $t \in \mathfrak{T}$, the convergence of*

$$\lim_{n\to\infty} R_k\big(\mathfrak{m}_{t(n)\sqrt{n}}\big) = R_k(\mathfrak{m}_\Omega), \qquad k \in \mathbb{N} \tag{4.8}$$

holds.

The rest of this section is devoted to the proof of Theorem 4.3. However, some explanatory comments are scattered as Remarks on the way.

We have $R_k(\mathfrak{m}_\Omega) = \delta_{k,2}$ by (3.15), $R_1(\mathfrak{m}_{\lambda\sqrt{n}}) = 0$ and $R_2(\mathfrak{m}_{\lambda\sqrt{n}}) = 1$ for $\lambda \in \mathbb{Y}_n$ by (2.19) and (2.11). Hence (4.8) is equivalent to

$$\lim_{n\to\infty} R_k\big(\mathfrak{m}_{t(n)\sqrt{n}}\big) = 0, \qquad k \in \mathbb{N}, \ k \geq 3. \tag{4.9}$$

It suffices to show that, for a fixed $k \geq 3$, (4.9) holds for M_{Pl}-a.s. $t \in \mathfrak{T}$.

Remark 4.1 As many textbooks on probability theory tell us, the strong law of large numbers for the sum of i.i.d. $\mathbb{R}$-valued random variables $\{X_n\}$ easily follows from the Borel–Cantelli lemma if X_n is assumed to have the fourth moment. Considering the expectation of $R_k(\mathfrak{m}_{t(n)\sqrt{n}})^4$ below is suggested by this fact.

For $c > 0$, let $\mathbb{Y}_{n,c}$ denote the subset of $\mathbb{Y}_n$ consisting of c-balanced Young diagrams (namely, satisfying (4.5)). As a result we will have

$$\sum_{n=1}^{\infty} \int_{\{t \in \mathfrak{T} \,|\, t(n) \in \mathbb{Y}_{n,c}\}} R_k(\mathfrak{m}_{t(n)\sqrt{n}})^4 M_{\mathrm{Pl}}(dt) < \infty \tag{4.10}$$

for any $c > 0$ and $k \geq 3$. First we verify that the desired consequence follows from (4.10).

Since (4.10) with the Chebyshev inequality gives

$$\sum_{n=1}^{\infty} M_{\mathrm{Pl}}\big(\{t \in \mathfrak{T} \,|\, t(n) \in \mathbb{Y}_{n,c}, \ |R_k(\mathfrak{m}_{t(n)\sqrt{n}})| \geq \varepsilon\}\big) < \infty$$

for any $\varepsilon > 0$, the Borel–Cantelli lemma yields that, for M_{Pl}-a.s. $t \in \mathfrak{T}$, $t(n) \notin \mathbb{Y}_{n,c}$ or $|R_k(\mathfrak{m}_{t(n)\sqrt{n}})| < \varepsilon$ holds except finite n's. On the other hand, Theorem 4.1 tells that, if $c > 0$ is large enough, then, for M_{Pl}-a.s. $t \in \mathfrak{T}$, $t(n) \in \mathbb{Y}_{n,c}$ holds except finite n's. Hence we have, for M_{Pl}-a.s. $t \in \mathfrak{T}$, $t(n) \in \mathbb{Y}_{n,c}$ and $|R_k(\mathfrak{m}_{t(n)\sqrt{n}})| < \varepsilon$ hold except finite n's. Taking a decreasing sequence $\varepsilon_j \downarrow 0$, set

$$\mathfrak{T}^{(0)} = \bigcap_{k=3}^{\infty} \bigcap_{j=1}^{\infty} \{t \in \mathfrak{T} \,|\, t(n) \in \mathbb{Y}_{n,c}, \ |R_k(\mathfrak{m}_{t(n)\sqrt{n}})| < \varepsilon_j \text{ except finite } n\text{'s}\}.$$

We have $M_{\mathrm{Pl}}(\mathfrak{T}^{(0)}) = 1$ and that (4.9) holds if $t \in \mathfrak{T}^{(0)}$.

Remark 4.2 Actually, the property of elements of $\mathfrak{T}^{(0)}$ assures that we have proved the convergence of Theorem 4.2 with respect to the inductive limit topology on $\mathbb{D}$ (see Remark 3.2).

Remark 4.3 Recall that $\{R_k(\mathfrak{m}_\lambda)\}_{k \in \{2,3,\dots\}}$ forms a generating set of the Kerov–Olshanski algebra $\mathbb{A}$. Among several generating sets of $\mathbb{A}$ mentioned in Sect. 2.3, $\Sigma_k(\lambda)$ of (2.21) (irreducible character of the symmetric group) is best fit to compute the expectation with respect to the Plancherel measure. We know the transition rules between generators of $\mathbb{A}$ to a more or less considerable extent.

To show (4.10), we apply Theorem 2.2 and replace R_k with Σ_{k-1}, taking Remark 4.3 into account. We have

$$\int_{\{t \in \mathfrak{T} \,|\, t(n) \in \mathbb{Y}_{n,c}\}} R_k(\mathfrak{m}_{t(n)\sqrt{n}})^4 M_{\mathrm{Pl}}(dt)$$

$$= \sum_{\lambda \in \mathbb{Y}_{n,c}} R_k(\mathfrak{m}_{\lambda\sqrt{n}})^4 M_{\mathrm{Pl}}^{(n)}(\{\lambda\}) = \sum_{\lambda \in \mathbb{Y}_{n,c}} n^{-2k} R_k(\mathfrak{m}_\lambda)^4 M_{\mathrm{Pl}}^{(n)}(\{\lambda\})$$

$$= \sum_{\lambda \in \mathbb{Y}_{n,c}} n^{-2k} \big\{\Sigma_{k-1}(\lambda) - P_{k-1}\big(R_2(\mathfrak{m}_\lambda), \dots, R_{k-2}(\mathfrak{m}_\lambda)\big)\big\}^4 M_{\mathrm{Pl}}^{(n)}(\{\lambda\})$$

$$\leqq 8 \sum_{\lambda \in \mathbb{Y}_{n,c}} n^{-2k} \Sigma_{k-1}(\lambda)^4 M_{\mathrm{Pl}}^{(n)}(\{\lambda\})$$

$$+ 8 \sum_{\lambda \in \mathbb{Y}_{n,c}} n^{-2k} P_{k-1}\big(R_2(\mathfrak{m}_\lambda), \ldots, R_{k-2}(\mathfrak{m}_\lambda)\big)^4 M_{\mathrm{Pl}}^{(n)}(\{\lambda\}), \qquad (4.11)$$

where note that Corollary 3.1 (with the free cumulant-moment formula) yields

$$R_k(\mathfrak{m}_{\lambda\sqrt{n}}) = n^{-k/2} R_k(\mathfrak{m}_\lambda), \qquad \lambda \in \mathbb{Y}_n.$$

The second sum of (4.11) is treated by easy weight-degree counting. In fact, we know by Corollary 3.1 that, if $\lambda \in \mathbb{Y}_{n,c}$, $\mathrm{supp}\, \mathfrak{m}_{\lambda\sqrt{n}} \subset [-c, c]$ and hence

$$|R_j(\mathfrak{m}_\lambda)| = n^{j/2} |R_j(\mathfrak{m}_{\lambda\sqrt{n}})| \leqq C' n^{j/2} \qquad (4.12)$$

for some $C' > 0$ depending only on c and j. In (4.11),

$$\mathrm{wt}\, P_{k-1}\big(R_2(\mathfrak{m}_\lambda), \ldots, R_{k-2}(\mathfrak{m}_\lambda)\big) \leqq k - 2 \qquad (4.13)$$

holds. Hence we have from (4.12) and $\mathrm{wt}\, R_j(\mathfrak{m}_\lambda) = j$

$$\text{2nd sum of(4.11)} \leqq C'' \sum_{\lambda \in \mathbb{Y}_{n,c}} n^{-2k} n^{4(k-2)/2} M_{\mathrm{Pl}}^{(n)}(\{\lambda\}) \leqq C'' n^{-4} \qquad (4.14)$$

for some $C'' > 0$ depending only on c and k.

Remark 4.4 As is seen below, it is sufficient for our present purpose that (4.14) holds for n^{-2} instead of n^{-4}, and so does (4.13) for $k - 1$ instead of $k - 2$.

To rewrite the first sum of (4.11), set $A_\rho = \sum_{x \in C_\rho} x$ for a conjugacy class C_ρ of $\mathfrak{S}_n$ (see (1.2)), which belongs to the center of $\mathbb{C}[\mathfrak{S}_n]$. Since a normalized irreducible character of $\mathfrak{S}_n$ is multiplicative on the center of $\mathbb{C}[\mathfrak{S}_n]$, we have

$$\sum_{\lambda \in \mathbb{Y}_n} n^{-2k} \Sigma_{k-1}(\lambda)^4 M_{\mathrm{Pl}}^{(n)}(\{\lambda\}) = \sum_{\lambda \in \mathbb{Y}_n} n^{-2k} (k-1)^4 \tilde{\chi}^\lambda\big(A_{(k-1,1^{n-k+1})}^4\big) M_{\mathrm{Pl}}^{(n)}(\{\lambda\})$$

$$= (k-1)^4 n^{-2k} \delta_e\big(A_{(k-1,1^{n-k+1})}^4\big). \qquad (4.15)$$

It is easy to see that

$$\delta_e\big(A_{(k-1,1^{n-k+1})}^4\big) = \sum_{w,x,y,z \in C_{(k-1,1^{n-k+1})}} \delta_e(wxyz) \qquad (4.16)$$

is $O(n^{2(k-1)})$ as $n \to \infty$. In fact, let $r = |(\text{supp}\, w) \cup (\text{supp}\, x) \cup (\text{supp}\, y) \cup (\text{supp}\, z)|$ in (4.16). If $2r > |\text{supp}\, w| + |\text{supp}\, x| + |\text{supp}\, y| + |\text{supp}\, z| = 4(k-1)$, $wxyz$ cannot coincide with e. Hence we have only to take into account w, x, y, z such that $r \leqq 2(k-1)$. Combining this with (4.15), we have verified that

$$\text{1st sum of (4.11)} \leqq C''' n^{-2k} n^{2(k-1)} = C''' n^{-2} \tag{4.17}$$

for some $C''' > 0$ depending only on k. Now (4.10) follows from (4.11), (4.17) and (4.14). This completes the proof of Theorem 4.3.

Remark 4.5 In the above proof of Theorem 4.3, we showed (4.10) to apply the Borel–Cantelli lemma with the help of Theorem 4.1. The balanced condition for λ's made the estimate for $P_{k-1}(R_2(\mathfrak{m}_\lambda), \ldots, R_{k-2}(\mathfrak{m}_\lambda))$ in (4.11) easy. On the other hand, without knowing the balancedness in advance, we can modify the proof by expressing the lower weight-degree terms by Σ_j's instead of R_j's.

For a modification of the proof of Theorem 4.3, in which the balanced condition is not explicit, we show

$$\sum_{n=1}^{\infty} \int_{\mathfrak{T}} R_k(\mathfrak{m}_{t(n)\sqrt{n}})^4 M_{\text{Pl}}(dt) < \infty, \qquad k \geq 3 \tag{4.18}$$

instead of (4.10). We use Theorem 2.2 to get, similarly to (4.11),

$$\int_{\mathfrak{T}} R_k(\mathfrak{m}_{t(n)\sqrt{n}})^4 M_{\text{Pl}}(dt) \leqq 8 \sum_{\lambda \in \mathbb{Y}_n} n^{-2k} \Sigma_{k-1}(\lambda)^4 M_{\text{Pl}}^{(n)}(\{\lambda\})$$

$$+ 8 \sum_{\lambda \in \mathbb{Y}_n} n^{-2k} Q\big(\Sigma_1(\lambda), \ldots, \Sigma_{k-3}(\lambda)\big)^4 M_{\text{Pl}}^{(n)}(\{\lambda\}) \tag{4.19}$$

where Q is a polynomial satisfying wt $Q\big(\Sigma_1(\lambda), \ldots, \Sigma_{k-3}(\lambda)\big) \leqq k - 2$. Note that $Q = 0$ for $k = 3$ since $R_3(\mathfrak{m}_\lambda) = \Sigma_2(\lambda)$ holds. The function $Q\big(\Sigma_1(\lambda), \ldots, \Sigma_{k-3}(\lambda)\big)$ is a linear combination of $\Sigma_{j_1}(\lambda)\Sigma_{j_2}(\lambda) \ldots \Sigma_{j_p}(\lambda)$'s, which satisfies

$$\text{wt } \Sigma_{j_1} \Sigma_{j_2} \ldots \Sigma_{j_p} = (j_1 + 1) + \cdots + (j_p + 1) = j_1 + \cdots + j_p + p \leqq k - 2. \tag{4.20}$$

To estimate (4.19), we again use multiplicativity of $\tilde{\chi}^\lambda$ on the center of $\mathbb{C}[\mathfrak{S}_n]$ with noticing, however, appearance of Σ_1. Let $j_1, \ldots, j_q \geq 2$ and $j_{q+1} = \cdots = j_p = 1$ for $0 \leq q \leq p$. Then, we have

$$\sum_{\lambda \in \mathbb{Y}_n} n^{-2k} \left(\Sigma_{j_1}(\lambda) \dots \Sigma_{j_p}(\lambda) \right)^4 M_{\mathrm{Pl}}^{(n)}(\{\lambda\})$$

$$= \sum_{\lambda \in \mathbb{Y}_n} n^{-2k+4(p-q)} \left(\Sigma_{j_1}(\lambda) \dots \Sigma_{j_q}(\lambda) \right)^4 M_{\mathrm{Pl}}^{(n)}(\{\lambda\})$$

$$= \sum_{\lambda \in \mathbb{Y}_n} n^{-2k+4(p-q)} j_1^4 \dots j_q^4 \tilde{\chi}^{\lambda} \left(\prod_{i=1}^{q} A_{(j_i, 1^{n-j_i})}^{4} \right) M_{\mathrm{Pl}}^{(n)}(\{\lambda\})$$

$$= n^{-2k+4(p-q)} j_1^4 \dots j_q^4 \delta_e \left(\prod_{i=1}^{q} A_{(j_i, 1^{n-j_i})}^{4} \right). \tag{4.21}$$

Similarly to the estimate for (4.16), we see

$$(4.21) \leqq K' n^{-2k+4(p-q)} n^{2(j_1+\cdots+j_q)} \leqq K' n^{-2k+4(p-q)+2(k-2-2p+q)} = K' n^{-4-2q}$$

by (4.20), where $K' > 0$ depends only on $j_1, \dots, j_p$. We have thus

$$\text{2nd sum of (4.19)} \leqq K n^{-4}$$

for some $K > 0$ depending only on k. This leads to the proof of (4.18), and hence of Theorem 4.3.

Remark 4.6 In [14], we included a (partly non-self-contained) proof of the limit shape for the Plancherel measure of symmetric groups in terms of analysis of the Jucys–Murphy elements. Actually, overlap of this book with [14] is not so much except presentations of basic notions on Young diagrams. In the latter, the main ingredient concerning asymptotic representation theory was the fluctuation (CLT), which we analyzed by the method of quantum decomposition therein.

4.3 Continuous Hook and the Limit Shape

In this section, we present a variational approach to the limit shape problem. We follow the discussion described in [19, 31] while a full detail of computation is left to [13]. We give a proof of Theorem 4.2 relying on the hook formula (Proposition 1.1), balancedness of Young diagrams (Theorem 4.1) and a famous formula of Hardy–Ramanujan (Theorem 4.4). As a result, group representation theory is not needed for the proof.

Theorem 4.4 *The number of Young diagrams of size n satisfies*

$$|\mathbb{Y}_n| = \frac{e^{\pi \sqrt{2n/3}}}{4\sqrt{3}n} \left(1 + O\left(\frac{1}{\sqrt{n}}\right) \right)$$

as $n \to \infty$.

Fig. 4.2 Continuous hook length

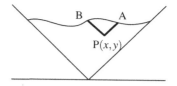

See [7] for the proof of Theorem 4.4.

We begin with introducing a continuous version of a hook length for a continuous diagram. For $\omega \in \mathbb{D}$, let $D(\omega)$ denote the region $\{(x, y) \in \mathbb{R}^2 \mid |x| < y < \omega(x)\}$. In particular, if $\lambda \in \mathbb{Y}_n$, the area of $D(\lambda)$ is $2n$. As indicated in Fig. 4.2, take P(x, y), A(s, ξ) and B(t, η) for $(x, y) \in D(\omega)$, and set $h_\omega(x, y) = $ PA+PB. From an obvious reason, $h_\omega(x, y)$ is called the continuous hook length at (x, y). We have

$$h_\omega(x, y) = \sqrt{2}(s - x) + \sqrt{2}(x - t) = \sqrt{2}(s - t), \qquad (x, y) \in D(\omega)$$

where (s, t) is the solution of

$$\begin{cases} \xi - y = s - x \\ \xi = \omega(s) \end{cases}, \qquad \begin{cases} \eta - y = -(t - x) \\ \eta = \omega(t). \end{cases} \tag{4.22}$$

Our first task is to show the following asymptotic formula for the Plancherel measure.

Proposition 4.2 *We have*

$$M_{\mathrm{Pl}}^{(n)}(\{\lambda\}) = \left(1 + o(1)\right)\sqrt{2\pi n}$$
$$\exp\left\{-n\left(1 + \iint_{D(\lambda^{\sqrt{n}})} \log \frac{h_{\lambda^{\sqrt{n}}}(x, y)}{\sqrt{2}} dx dy + O\left(\frac{1}{\sqrt{n}}\right)\right)\right\} \tag{4.23}$$

for $\lambda \in \mathbb{Y}_n$ as $n \to \infty$. The two error terms in (4.23) depend only on n.

Proof [*Step* 1] Taking it into account that $\lambda \in \mathbb{Y}_n$ is rescaled by $1/\sqrt{n}$, we rewrite $M_{\mathrm{Pl}}^{(n)}(\{\lambda\})$ by using Proposition 1.1 and Stirling's formula as

$$M_{\mathrm{Pl}}^{(n)}(\{\lambda\}) = n! \exp\left(-2\sum_{b \in \lambda} \log h_\lambda(b)\right) = \frac{n! 2^n}{n^n} \exp\left(-n\sum_{b \in \lambda} \frac{2}{n} \log \frac{\sqrt{2}h_\lambda(b)}{\sqrt{n}}\right)$$
$$= \left(1 + o(1)\right)\sqrt{2\pi n} \exp\left\{-n\left(1 - \log 2 + \sum_{b \in \lambda} \frac{2}{n} \log \frac{\sqrt{2}h_\lambda(b)}{\sqrt{n}}\right)\right\}. \tag{4.24}$$

For $b \in \lambda \in \mathbb{Y}_n$, we replace $h_{\lambda^{\sqrt{n}}}(x, y)$, the continuous hook length at $(x, y) \in b^{\sqrt{n}}$ (a small box in $\lambda^{\sqrt{n}}$), with the value at the center (x_0, y_0) of $b^{\sqrt{n}}$ to have

$$\tilde{h}_{\lambda\sqrt{n}}(x, y) = h_{\lambda\sqrt{n}}(x_0, y_0) = \sqrt{\frac{2}{n}}h_\lambda(b), \qquad (x, y) \in b^{\sqrt{n}}. \qquad (4.25)$$

Putting (4.25) into (4.24), we have

$$M_{\mathrm{Pl}}^{(n)}(\{\lambda\}) = (1 + o(1))\sqrt{2\pi n}\exp\left\{-n\left(1 + \iint_{D(\lambda\sqrt{n})}\log\frac{\tilde{h}_{\lambda\sqrt{n}}(x, y)}{\sqrt{2}}dxdy\right)\right\}.$$

[*Step* 2] We show

$$\iint_{D(\lambda\sqrt{n})}\left(\log h_{\lambda\sqrt{n}}(x, y) - \log\tilde{h}_{\lambda\sqrt{n}}(x, y)\right)dxdy = O\left(\frac{1}{\sqrt{n}}\right) \qquad (4.26)$$

holds as $n \to \infty$. First we write the integral on a small box $b^{\sqrt{n}}$, which equals

$$\int_{-1/\sqrt{2n}}^{1/\sqrt{2n}}\int_{-1/\sqrt{2n}}^{1/\sqrt{2n}}\left(\log(h - x - y) - \log h\right)dxdy$$
$$= \frac{h^2}{2}\left(1 + \frac{\sqrt{2}}{h\sqrt{n}}\right)^2\log\left(1 + \frac{\sqrt{2}}{h\sqrt{n}}\right) + \frac{h^2}{2}\left(1 - \frac{\sqrt{2}}{h\sqrt{n}}\right)^2\log\left(1 - \frac{\sqrt{2}}{h\sqrt{n}}\right) - \frac{3}{n}$$

where h is the continuous hook length at the center of $b^{\sqrt{n}}$. Since (4.25) yields $h\sqrt{n/2} = h_\lambda(b)$, we continue as

$$= \frac{1}{n}\left\{(h_\lambda(b) + 1)^2\log\left(1 + \frac{1}{h_\lambda(b)}\right) + (h_\lambda(b) - 1)^2\log\left(1 - \frac{1}{h_\lambda(b)}\right) - 3\right\}. \quad (4.27)$$

Note that (4.27) is valid also if $h_\lambda(b) = 1$. Now sum up (4.27) over all boxes b in λ to get

$$\left|\text{LHS of (4.26)}\right| = \sum_{j=1}^n \left|\{b \in \lambda \mid h_\lambda(b) = j\}\right|$$
$$\times \frac{1}{n}\left|(j + 1)^2\log\left(1 + \frac{1}{j}\right) + (j - 1)^2\log\left(1 - \frac{1}{j}\right) - 3\right|. \quad (4.28)$$

In (4.28), we have

$$(j + 1)^2\log\left(1 + \frac{1}{j}\right) + (j - 1)^2\log\left(1 - \frac{1}{j}\right) - 3$$
$$= \sum_{k=1}^\infty\left(-\frac{1}{k} - \frac{1}{k + 1} + \frac{4}{2k + 1}\right)\frac{1}{j^{2k}} = -\sum_{k=1}^\infty\frac{1}{k(k + 1)(2k + 1)j^{2k}}$$

for $j \geq 2$ (and <0 for $j = 1$ also). Using this with (4.29) below in [Step 3], we have

$$(4.28) \leq \frac{C'}{\sqrt{n}} + \sum_{j=2}^{n} \sum_{k=1}^{\infty} \frac{\sqrt{2}}{\sqrt{n}} \frac{1}{k(k+1)(2k+1)j^{2k-(1/2)}} \leq \frac{C}{\sqrt{n}}$$

for some positive constants C' and C.

[*Step* 3] To complete the proof, we show

$$\max_{\lambda \in \mathbb{Y}_n} |\{b \in \lambda \mid h_\lambda(b) = j\}| \leq \sqrt{2jn}, \quad 1 \leq j \leq n. \qquad (4.29)$$

Imagine the English display of a Young diagram. Let $\tilde{b}$ denote the box which the hook at box b shares with the horizontal strip of the border of λ, in other words, the end of the leg from b. If b itself lies in a horizontal strip of the border, we set $\tilde{b} = b$. If $\tilde{b}$ is distant by more than j from the right-nearest corner, the hook at b does not have length j. For each horizontal strip of the border, therefore, we have only to consider at most ($j \wedge$ (the horizontal strip length)) hooks. Hence the number of the hooks of length j is bounded above by

$$u = \sum_{i:\, m_i(\lambda') \geq 1} j \wedge m_i(\lambda').$$

For given n and j, the quantity u is maximized by $\lambda \in \mathbb{Y}_n$ as indicated in Fig. 4.3. Letting the border of λ in Fig. 4.3 contains $(p-1)$ horizontal strips of length j, we have

$$n = \frac{1}{2}jp(p-1) + j'p, \quad u = j(p-1) + j', \quad 0 \leq j' \leq j-1. \qquad (4.30)$$

Then, $u^2 \leq 2jn$ follows from (4.30). Indeed,

$$2jn - u^2 = j^2(p-1) + j'(2j - j') \geq 0.$$

We thus obtain (4.29).

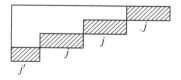

(p − 1) horizontal strips of length j

possibly 1 strip of length $j' < j$

Fig. 4.3 Young diagram in Step 3 of the proof

We consider the change of variables through (4.22) i.e.

$$x = \frac{1}{2}\big(s - \omega(s) + t + \omega(t)\big), \quad y = \frac{1}{2}\big(t + \omega(t) - s + \omega(s)\big) \tag{4.31}$$

for the integral in (4.23), or more generally for $\omega \in \mathbb{D}$. The map $(s, t) \mapsto (x, y)$ by (4.31) is surjective with the Jacobian

$$\frac{\partial x}{\partial s}\frac{\partial y}{\partial t} - \frac{\partial x}{\partial t}\frac{\partial y}{\partial s} = \frac{1}{2}\big(1 - \omega'(s)\big)\big(1 + \omega'(t)\big). \tag{4.32}$$

Moreover, it is bijective if restricted on the region where (4.32) does not vanish. We thus have

$$\iint_{D(\omega)} \log \frac{h_\omega(x, y)}{\sqrt{2}} dx dy = \frac{1}{2} \iint_{\{s>t\}} \big(1 - \omega'(s)\big)\big(1 + \omega'(t)\big) \log(s - t) ds dt. \tag{4.33}$$

Comparing (4.33) with (4.23), we set

$$\theta(\omega) = 1 + \frac{1}{2} \iint_{\{s>t\}} \big(1 - \omega'(s)\big)\big(1 + \omega'(t)\big) \log(s - t) ds dt, \quad \omega \in \mathbb{D}. \tag{4.34}$$

Proposition 4.3 *The functional θ on $\mathbb{D}$ satisfies the following:*

$$\theta(\Omega) = 0, \tag{4.35}$$

$$\theta(\Omega + \phi) = \frac{1}{4} \iint_{\{s>t\}} \Big(\frac{\phi(s) - \phi(t)}{s - t}\Big)^2 ds dt + \int_{\{|s|\geq 2\}} \phi(s) \log\Big(\frac{|s|}{2} + \sqrt{\frac{s^2}{4} - 1}\Big) ds \tag{4.36}$$

if $\Omega + \phi \in \mathbb{D}$, and furthermore Ω is a unique minimizer for θ on $\mathbb{D}$.

Proof We show an outline of the computation. See also (4.39) and (4.40). Starting from (4.34), we have the following after some computation:

$$\theta(\Omega) = 1 + \frac{1}{2} \iint_{\{-2<t<s<2\}} \Big(1 - \frac{2}{\pi}\arcsin\frac{s}{2}\Big)\Big(1 + \frac{2}{\pi}\arcsin\frac{t}{2}\Big) \log(s - t) ds dt$$

$$= 1 + \frac{1}{2} \int_{-2}^{2} \Big\{\int_t^2 \Big(1 - \frac{2}{\pi}\arcsin\frac{s}{2}\Big) \log(s - t) ds\Big\} dt$$

$$= 1 + \int_0^2 (s \log s - s)\Big(1 - \frac{2}{\pi}\arcsin\frac{s}{2}\Big) ds = 0.$$

Using (4.34) and (4.35), we have

$$\theta(\Omega + \phi) = \frac{1}{2}\int_{-\infty}^{-2}\left(\int_{t}^{2}(1 - \Omega'(s))\log(s - t)ds\right)\phi'(t)dt$$
$$- \frac{1}{2}\int_{2}^{\infty}\left(\int_{-2}^{s}(1 + \Omega'(t))\log(s - t)dt\right)\phi'(s)ds$$
$$- \frac{1}{2}\iint_{\{s>t\}}\phi'(s)\phi'(t)\log(s - t)dsdt. \qquad (4.37)$$

The third term of (4.37) equals the first of (4.36). Indeed, take $a > 0$ such that $\operatorname{supp}\phi \subset (-a, a)$ to have

$$-\frac{1}{2}\iint_{\{s>t\}}\phi'(s)\phi'(t)\log(s - t)dsdt$$
$$= \frac{a}{2}\int_{-a}^{a}\frac{\phi(s)^2}{a^2 - s^2}ds + \frac{1}{4}\iint_{\{-a<t<s<a\}}\left(\frac{\phi(s) - \phi(t)}{s - t}\right)^2 dsdt \quad (4.38)$$

and let $a \to \infty$. The first and second terms of (4.37) give the second of (4.36).

Finally, since $\Omega + \phi \in \mathbb{D}$ implies that $\phi(s) \geq 0$ for $|s| \geq 2$, (4.36) yields that Ω is a unique minimizer of θ.

We mention some integrations for reference: for $s \in \mathbb{R}$,

$$\int_{-2}^{2}\frac{\log|s - x|}{\pi\sqrt{4 - x^2}}dx = \begin{cases} \log\left(\frac{|s|}{2} + \sqrt{\frac{s^2}{4} - 1}\right), & |s| \geq 2, \\ 0, & |s| \leq 2, \end{cases} \qquad (4.39)$$

$$\int_{-2}^{2}\frac{x\log|s - x|}{\pi\sqrt{4 - x^2}}dx = \begin{cases} -s - \sqrt{s^2 - 4}, & s \leq -2, \\ -s, & -2 \leq s \leq 2, \\ -s + \sqrt{s^2 - 4}, & 2 \leq s. \end{cases} \qquad (4.40)$$

Putting (4.23), (4.33) and (4.34) together, we get the asymptotic

$$M_{\mathrm{Pl}}^{(n)}(\{\lambda\}) = (1 + o(1))\sqrt{2\pi n}\, e^{-n\theta(\lambda^{\sqrt{n}}) + O(\sqrt{n})}, \qquad \lambda \in \mathbb{Y}_n \qquad (4.41)$$

as $n \to \infty$. Combining (4.41) with Theorem 4.4, we obtain the following.

Theorem 4.5 *For any $\varepsilon > 0$,*

$$M_{\mathrm{Pl}}^{(n)}\left(\{\lambda \in \mathbb{Y}_n \mid \theta(\lambda^{\sqrt{n}}) \geq \varepsilon\}\right) \leq \frac{C_1}{\sqrt{n}}\, e^{-\varepsilon n + C_2\sqrt{n}} \qquad (4.42)$$

holds with universal positive constants C_1 and C_2.

Let us now give an alternative proof of Theorem 4.2 by applying Theorem 4.5. We can take $c > 2$ and $\mathfrak{T}_0 = \{t \in \mathfrak{T} \mid t(n) \in \mathbb{Y}_{n,c}$ for sufficiently large $n\}$ such that

$M_{\mathrm{Pl}}(\mathfrak{T}_0) = 1$ by Theorem 4.1. Note that $\lambda \in \mathbb{Y}_{n,c}$ satisfies $\mathrm{supp}(\lambda^{\sqrt{n}} - \Omega) \subset [-c, c]$ since $c > 2$. Then, (4.37) and (4.38) yield

$$\theta(\lambda^{\sqrt{n}}) = \theta(\Omega + \phi) \geqq \frac{c}{2} \int_{-c}^{c} \frac{\phi(s)^2}{c^2 - s^2} ds, \qquad \phi = \lambda^{\sqrt{n}} - \Omega. \qquad (4.43)$$

Considering an estimate from below for $\phi(s)$ in (4.43) by a suitable triangle, we see that, for $\lambda \in \mathbb{Y}_{n,c}$,

$$\|\lambda^{\sqrt{n}} - \Omega\|_{\sup} \geqq \varepsilon \implies \theta(\lambda^{\sqrt{n}}) \geqq \varepsilon^3/(6c).$$

Therefore, it follows from (4.42) that

$$M_{\mathrm{Pl}}^{(n)}\left(\{\lambda \in \mathbb{Y}_{n,c} \mid \|\lambda^{\sqrt{n}} - \Omega\|_{\sup} \geqq \varepsilon\}\right) \leqq \frac{C_1}{\sqrt{n}} \exp\left(-\frac{\varepsilon^3 n}{6c} + C_2\sqrt{n}\right).$$

By the Borel–Cantelli lemma, we have $\mathfrak{T}_1 \subset \mathfrak{T}$ such that $M_{\mathrm{Pl}}(\mathfrak{T}_1) = 1$ and

$$t \in \mathfrak{T}_1 \implies t(n) \notin \mathbb{Y}_{n,c} \text{ or } \|t(n)^{\sqrt{n}} - \Omega\|_{\sup} < \varepsilon \text{ with finite exceptional } n\text{'s.}$$

Hence $t \in \mathfrak{T}_0 \cap \mathfrak{T}_1$ satisfies $\|t(n)^{\sqrt{n}} - \Omega\|_{\sup} < \varepsilon$ for sufficiently large n. Routinely, replace ε by $\varepsilon_j \downarrow 0$. This completes the proof of Theorem 4.2.

4.4 Approximate Factorization Property

Although this book is mainly concerned with the Plancherel measure, in this section we look into some progress of the limit shape problem observed in other random structures. The key notion here is the approximate factorization property of a state of the group algebra, introduced by Biane in [2], which nicely weakens the ergodicity of a measure. Another motivation for touching upon a bit general theory here is to help to see a variety of initial states for the dynamical model treated in Chap. 5.

We begin with setting a condition for a sequence of probability spaces to formulate the concentration of profiles. For the sake of convenience, let us say that a sequence of probability space $\{(\mathbb{Y}_n, \mathbb{M}^{(n)})\}_{n \in \mathbb{N}}$ admits the concentration at ψ if there exists $\psi \in \mathbb{D}$ such that, for any $p \in \mathbb{N}$ and $k_1, \ldots, k_p \in \{2, 3, \ldots\}$,

$$\lim_{n \to \infty} E_{\mathbb{M}^{(n)}}\left[M_{k_1}(\mathfrak{m}_{\lambda^{\sqrt{n}}}) \ldots M_{k_p}(\mathfrak{m}_{\lambda^{\sqrt{n}}})\right] = M_{k_1}(\mathfrak{m}_{\psi}) \ldots M_{k_p}(\mathfrak{m}_{\psi}) \qquad (4.44)$$

holds. The condition (4.44) uniquely determines ψ. Furthermore, (4.44) holds automatically for $k_i = 0$ or 1. It is obvious that (4.44) yields

$$\lim_{n \to \infty} E_{\mathbb{M}^{(n)}}\left[\left(M_k(\mathfrak{m}_{\lambda^{\sqrt{n}}}) - M_k(\mathfrak{m}_{\psi})\right)^2\right] = 0, \qquad k \in \mathbb{N} \cup \{0\}$$

and hence the weak law of large numbers with respect to the moment topology on $\mathbb{D}$.

The following two conditions are an equivalent modification of (4.44): there exist a real sequence $\{m_3, m_4, \ldots\}$ and $a > 0$ such that

$$\lim_{n \to \infty} E_{\mathbb{M}^{(n)}}\left[M_{k_1}(\mathfrak{m}_{\lambda\sqrt{n}}) \ldots M_{k_p}(\mathfrak{m}_{\lambda\sqrt{n}})\right] = m_{k_1} \ldots m_{k_p}, \tag{4.45}$$

$$|m_k| \leqq a^k. \tag{4.46}$$

In fact, (4.45) and (4.46) follow from (4.44). Conversely, if we set $m_0 = m_2 = 1$ and $m_1 = 0$, (4.45) holds for $k_i = 0, 1, 2$. Then, (4.45) and (4.46) give $\mu \in \mathcal{P}(\mathbb{R})$ with compact support such that $M_k(\mu) = m_k$ ($k \in \mathbb{N} \cup \{0\}$). We obtain $\psi \in \mathbb{D}$ such that $\mathfrak{m}_\psi = \mu$ by virtue of (3.12) and (3.13).

Moreover, (4.45) and (4.46) are equivalent also to that there exist a real sequence $\{r_3, r_4, \ldots\}$ and $b > 0$ such that

$$\lim_{n \to \infty} E_{\mathbb{M}^{(n)}}\left[R_{k_1}(\mathfrak{m}_{\lambda\sqrt{n}}) \ldots R_{k_p}(\mathfrak{m}_{\lambda\sqrt{n}})\right] = r_{k_1} \ldots r_{k_p}, \tag{4.47}$$

$$|r_k| \leqq b^k \tag{4.48}$$

as seen from the free cumulant-moment formula by setting $r_1 = 0$ and $r_2 = 1$. Note a rough estimate of $|NC(k)| \leqq 4^k$. Under (4.47) and (4.48), $\{(\mathbb{Y}_n, \mathbb{M}^{(n)})\}_{n \in \mathbb{N}}$ admits the concentration at $\psi \in \mathbb{D}$ such that $R_k(\mathfrak{m}_\psi) = r_k$.

Lemma 4.3 *For a sequence of probability spaces* $\{(\mathbb{Y}_n, \mathbb{M}^{(n)})\}_{n \in \mathbb{N}}$, (4.47) *is equivalent to that: for* $j, j_i \in \{2, 3, \ldots\}$,

$$E_{\mathbb{M}^{(n)}}\left[\Sigma_{j_1} \ldots \Sigma_{j_p}\right] = E_{\mathbb{M}^{(n)}}\left[\Sigma_{j_1}\right] \ldots E_{\mathbb{M}^{(n)}}\left[\Sigma_{j_p}\right] + o\left(n^{(j_1 + \cdots + j_p + p)/2}\right) \quad (n \to \infty), \tag{4.49}$$

$$\lim_{n \to \infty} n^{-(j+1)/2} E_{\mathbb{M}^{(n)}}\left[\Sigma_j\right] = r_{j+1}. \tag{4.50}$$

(Note that (4.49) and (4.50) are valid when j or j_i equals 1 also by setting $r_2 = 1$.)

Proof The assertions follow from Theorem 2.2 with simple weight counting. For example, in

$$E_{\mathbb{M}^{(n)}}\left[\Sigma_{j_1} \ldots \Sigma_{j_p}\right] = E_{\mathbb{M}^{(n)}}\left[R_{j_1+1}(\mathfrak{m}_\lambda) \ldots R_{j_p+1}(\mathfrak{m}_\lambda)\right] + E_{\mathbb{M}^{(n)}}\left[Q\right],$$

$Q \in \mathbb{A}$ is expressed as a polynomial of $R_i(\mathfrak{m}_\lambda)$'s with wt $Q \leqq j_1 + \cdots + j_p + p - 2$.

Proposition 2.7 with weight counting yields the following.

Lemma 4.4 *For a sequence of probability spaces* $\{(\mathbb{Y}_n, \mathbb{M}^{(n)})\}_{n \in \mathbb{N}}$, (4.49) *and* (4.50) *are equivalent to*

$$E_{\mathbb{M}^{(n)}}[\Sigma_{\rho \sqcup \sigma}] - E_{\mathbb{M}^{(n)}}[\Sigma_\rho] E_{\mathbb{M}^{(n)}}[\Sigma_\sigma] = o\big(n^{(|\rho|+l(\rho)+|\sigma|+l(\sigma))/2}\big) \tag{4.51}$$

as $n \to \infty$ for any $\rho, \sigma \in \mathbb{Y}^\times$ and (4.50).

An element $f \in \mathcal{K}(\mathfrak{S}_n)$ is linearly extended to a tracial state of $\mathbb{C}[\mathfrak{S}_n]$, denoted by the same symbol f. For a probability space $(\mathbb{Y}_n, \mathbb{M}^{(n)})$, let $f^{(n)}$ be the tracial state of $\mathbb{C}[\mathfrak{S}_n]$ assigned to $\mathbb{M}^{(n)}$ through (1.5).

Lemma 4.5 *For a sequence of probability spaces* $\{(\mathbb{Y}_n, \mathbb{M}^{(n)})\}_{n\in\mathbb{N}}$, (4.51) *and* (4.50) *are equivalent to*

$$f^{(n)}_{(\rho\sqcup\sigma, 1^{n-|\rho|-|\sigma|})} - f^{(n)}_{(\rho, 1^{n-|\rho|})} f^{(n)}_{(\sigma, 1^{n-|\sigma|})} = o\big(n^{-(|\rho|-l(\rho)+|\sigma|-l(\sigma))/2}\big) \tag{4.52}$$

as $n \to \infty$ for any $\rho, \sigma \in \mathbb{Y}^\times$, and

$$\lim_{n\to\infty} n^{(j-1)/2} f^{(n)}_{(j,1^{n-j})} = r_{j+1}, \qquad j \in \{2, 3, \ldots\}. \tag{4.53}$$

(The case of $j = 1$ is valid under $r_2 = 1$.)

Lemma 4.5 is shown also through a weight counting argument. In the proofs of these lemmas, we note that either (4.49) + (4.50), (4.51) + (4.50) or (4.52) + (4.53) yields

$$E_{\mathbb{M}^{(n)}}[\Sigma_\rho] = O\big(n^{(|\rho|+l(\rho))/2}\big), \quad f^{(n)}_{(\rho, 1^{n-|\rho|})} = O\big(n^{-(|\rho|-l(\rho))/2}\big)$$

as $n \to \infty$ for $\rho \in \mathbb{Y}^\times$ (then trivially extended to $\rho \in \mathbb{Y}$).

Among the above conditions concerning the concentration for a sequence of probability spaces $\{(\mathbb{Y}_n, \mathbb{M}^{(n)})\}_{n\in\mathbb{N}}$, we call (4.52) the *approximate factorization property* after [2].

Example 4.1 Let $\lambda^{(n)} \in \mathbb{Y}_n$ be c-balanced for some $c > 0$ and assume that there exists $\psi \in \mathbb{D}$ such that $\lambda^{(n)\sqrt{n}}$ converges to ψ in $\mathbb{D}$ (with respect to the moment topology or the uniform one). It is obvious that $\{(\mathbb{Y}_n, \delta_{\lambda^{(n)}})\}_{n\in\mathbb{N}}$ satisfies (4.44) and hence admits the concentration at ψ. Note that, for any $\psi \in \mathbb{D}$ given, we can take such $c > 0$ and $\lambda^{(n)} \in \mathbb{Y}_{n,c}$.

Example 4.2 The Littlewood–Richardson measure $\mathbb{M}^{(\mu,\nu)}$ is associated with the outer product $\mu \circ \nu$ of $\mu \in \mathbb{Y}_m$ and $\nu \in \mathbb{Y}_n$ through (1.5):

$$\tilde{\chi}^{\mu\circ\nu} = \sum_{\lambda\in\mathbb{Y}_{m+n}} \mathbb{M}^{(\mu,\nu)}(\{\lambda\}) \tilde{\chi}^\lambda. \tag{4.54}$$

By using the Littlewood–Richardson coefficients $c^\lambda_{\mu\nu}$, (4.54) is rewritten as

$$\mathbb{M}^{(\mu,\nu)}(\{\lambda\}) = \frac{m!n!}{(m+n)!} \frac{c^\lambda_{\mu\nu} \dim\lambda}{\dim\mu \dim\nu}, \qquad \lambda \in \mathbb{Y}_{m+n}.$$

Let us take two sequences of c-balanced Young diagrams $\{\mu^{(m)} \in \mathbb{Y}_m\}_{m \in \mathbb{N}}$ and $\{\nu^{(n)} \in \mathbb{Y}_n\}_{n \in \mathbb{N}}$ for some $c > 0$ whose rescaled profiles converge to $\phi \in \mathbb{D}^{(c)}$ and $\psi \in \mathbb{D}^{(c)}$ respectively, namely

$$\lim_{m \to \infty} (\mu^{(m)})^{\sqrt{m}} = \phi, \quad \lim_{n \to \infty} (\nu^{(n)})^{\sqrt{n}} = \psi \quad \text{in} \quad \mathbb{D}^{(c)} \subset \mathbb{D}.$$

Then, $\left\{ (\mathbb{Y}_{m+n}, \mathbb{M}^{(\mu^{(m)}, \nu^{(n)})}) \right\}_{m,n \in \mathbb{N}}$ admits the concentration at $\omega \in \mathbb{D}$ when $m, n \to \infty$ and $m/(m+n) \to q \in [0, 1]$. The limit profile ω is characterized by

$$\mathfrak{m}_\omega = \mathfrak{m}_{\phi^{1/\sqrt{q}}} \boxplus \mathfrak{m}_{\psi^{1/\sqrt{1-q}}} \tag{4.55}$$

where $\phi^{1/\sqrt{q}}(x) = \sqrt{q}\phi(x/\sqrt{q})$ similarly to (4.7). In fact, we have

$$\tilde{\chi}^{\mu^{(m)} \circ \nu^{(n)}}_{(k,1^{m+n-k})} = \frac{m^{\downarrow k}}{(m+n)^{\downarrow k}} \tilde{\chi}^{\mu^{(m)}}_{(k,1^{m-k})} + \frac{n^{\downarrow k}}{(m+n)^{\downarrow k}} \tilde{\chi}^{\nu^{(n)}}_{(k,1^{n-k})}, \quad k \geqq 2. \tag{4.56}$$

Theorem 2.2 transforms (4.56) into the asymptotic relation between free cumulants of the transition measures, which produces the free convolution in (4.55). The approximate factorization property follows also from the structure of induced representations. This fact of concentration for the Littlewood–Richardson measure was first obtained by Biane [1].

Example 4.3 Recall that $f \in \mathscr{K}(\mathfrak{S}_\infty)$ is extremal if and only if it is multiplicative as described in Theorem 1.3. Hence, if f is a character of $\mathfrak{S}_\infty$, $f^{(n)} = f\big|_{\mathfrak{S}_n}$ satisfies (4.52) without error terms. Let $(\alpha, \beta) \in \Delta$ correspond to $f = f_{\alpha,\beta}$ in Theorem 1.4. As shown by Vershik–Kerov [30], row and column lengths of the typical $t(n) \in \mathbb{Y}_n$ ($t \in \mathfrak{T}$) with respect to f are $n\alpha_i$ and $n\beta_i$ respectively as $n \to \infty$. In order to consider a macroscopic shape under the rescale by $1/\sqrt{n}$, we therefore adjust the Thoma parameter $(\alpha^{(n)}, \beta^{(n)}) \in \Delta$ to satisfy

$$\alpha_1^{(n)} = O(1/\sqrt{n}), \quad \beta_1^{(n)} = O(1/\sqrt{n}) \quad (n \to \infty), \tag{4.57}$$

and consider a sequence of probability spaces $\{(\mathbb{Y}_n, \mathbb{M}^{(n)})\}_{n \in \mathbb{N}}$ by taking $\mathbb{M}^{(n)} = \mathbb{M}^{(n)}_{\alpha^{(n)}, \beta^{(n)}}$ determined through (1.5):

$$f^{(n)} = f_{\alpha^{(n)}, \beta^{(n)}}\big|_{\mathfrak{S}_n} = \sum_{\lambda \in \mathbb{Y}_n} \mathbb{M}^{(n)}_{\alpha^{(n)}, \beta^{(n)}}(\{\lambda\}) \tilde{\chi}^\lambda.$$

For $(\alpha, \beta) \in \Delta$, set $\gamma = 1 - \sum_{i=1}^\infty (\alpha_i + \beta_i) \in [0, 1]$ and

$$\nu_{\alpha,\beta} = \sum_{i=1}^\infty (\alpha_i \delta_{\alpha_i} + \beta_i \delta_{-\beta_i}) + \gamma \delta_0 \in \mathscr{P}(\mathbb{R}).$$

Then, since

$$M_k(\nu_{\alpha,\beta}) = \sum_{i=1}^{\infty}(\alpha_i^{k+1} + (-1)^k\beta_i^{k+1}) = f_{\alpha,\beta}((k+1)\text{-cycle})$$

holds for $k \in \mathbb{N}$, we have

$$M_{j-1}\left(\nu_{\alpha^{(n)},\beta^{(n)}}\left(\frac{1}{\sqrt{n}}\cdot\right)\right) = n^{(j-1)/2}f^{(n)}_{(j,n-j)}, \qquad j \in \mathbb{N}.$$

The supports of $\nu_{\alpha^{(n)},\beta^{(n)}}\left(\frac{1}{\sqrt{n}}\cdot\right)$ ($n \in \mathbb{N}$) are uniformly bounded under (4.57). Consequently, if we take a sequence of Thoma parameters $\{(\alpha^{(n)}, \beta^{(n)})\}_{n\in\mathbb{N}}$ satisfying (4.57) and the condition that

$$\nu_{\alpha^{(n)},\beta^{(n)}}\left(\frac{1}{\sqrt{n}}\cdot\right) \text{ converges weakly to } \nu \text{ in } \mathcal{P}(\mathbb{R}),$$

then $\left\{(\mathbb{Y}_n, \mathbb{M}^{(n)}_{\alpha^{(n)},\beta^{(n)}})\right\}_{n\in\mathbb{N}}$ admits the concentration at $\omega = \omega_\nu$ such that

$$R_k(\mathfrak{m}_\omega) = \lim_{n\to\infty} M_{k-2}\left(\nu_{\alpha^{(n)},\beta^{(n)}}\left(\frac{1}{\sqrt{n}}\cdot\right)\right) = M_{k-2}(\nu), \qquad k \in \{2, 3, \ldots\}.$$

The R-transform of $\mathfrak{m}_\omega$ is given by

$$R_{\mathfrak{m}_\omega}(\zeta) = \sum_{k=2}^{\infty} R_k(\mathfrak{m}_\omega)\zeta^{k-1} = \int_{\mathbb{R}} \frac{\zeta}{1-\zeta x}\nu(dx),$$

which serves concrete computation of the limit profile ω. Some details and further aspects are found in [2, 4].

Remark 4.7 Beyond the concentration of profiles of Young diagrams, one naturally gets interested in fluctuation from the limit profile. A fundamental reference in this line with respect to the Plancherel measure is [17]. Further studies are found in [14, 16, 19, 26].

Chapter 5
Dynamic Model

Abstract In this chapter, we discuss a dynamical aspect of the limit shape problem for random Young diagrams. In a microscopic point of view, a continuous time Markov chain is introduced on the Young diagrams of size n which keeps the Plancherel measure invariant and has an initial distribution admitting the concentration at a profile as n tends to ∞. Our model is built on such a canonical setting. By considering a diffusive scaling limit in time versus space, we derive a macroscopic time evolution of the limit profile. The resulting evolution is described through the Kerov transition measure in terms of free-probabilistic notions.

5.1 Restriction-Induction Chain

We consider a Markov chain on $\mathbb{Y}_n$ as follows. For a given $\lambda \in \mathbb{Y}_n$, imagine its profile. Remove a box from one of its peaks according to a certain rate (which is connected with the Plancherel measure). We have $\xi \in \mathbb{Y}_{n-1}$ such that $\xi \nearrow \lambda$. Next put a box at one of the valleys of ξ according to a certain rate to have $\mu \in \mathbb{Y}_n$ such that $\xi \nearrow \mu$. The chain gets a transition from λ to μ in one step. Alternatively, we can first put a box and next remove a box in a similar way. If (1.4) and (1.9) are recalled, this chain is clearly produced by restriction and induction (alternatively, induction and restriction) for irreducible representations of symmetric groups. Let us here look at such a restriction-induction chain in a bit general setting.

Let G be a finite group and H its subgroup. Setting

$$c_{\lambda,\xi} = \left[\operatorname{Res}_H^G \lambda : \xi\right] = \left[\operatorname{Ind}_H^G \xi : \lambda\right]$$

for $\lambda \in \widehat{G}$ and $\xi \in \widehat{H}$, we have

$$\operatorname{Res}_H^G \lambda \cong \bigoplus_{\xi \in \widehat{H}} [c_{\lambda,\xi}]\xi, \qquad \operatorname{Ind}_H^G \xi \cong \bigoplus_{\lambda \in \widehat{G}} [c_{\lambda,\xi}]\lambda$$

hence

© The Author(s) 2016　　　　　　　　　　　　　　　　　　　　　　　　　　61
A. Hora, *The Limit Shape Problem for Ensembles of Young Diagrams*,
SpringerBriefs in Mathematical Physics, DOI 10.1007/978-4-431-56487-4_5

$$\mathrm{Ind}_H^G \mathrm{Res}_H^G \lambda \;\cong\; \bigoplus_{\mu \in \widehat{G}} \Big[\sum_{\xi \in \widehat{H}} c_{\lambda,\xi} c_{\mu,\xi} \Big] \mu, \qquad \lambda \in \widehat{G}. \tag{5.1}$$

Taking the dimension of (5.1), we obtain a transition probability

$$P_{\lambda\mu} = \frac{\dim \mu}{[G:H]\dim \lambda} \sum_{\xi \in \widehat{H}} c_{\lambda,\xi} c_{\mu,\xi}, \qquad \lambda, \mu \in \widehat{G}. \tag{5.2}$$

The Plancherel measure on $\widehat{G}$ is defined by

$$M_{\mathrm{Pl}}^G(\{\lambda\}) = (\dim \lambda)^2/|G|, \qquad \lambda \in \widehat{G}.$$

Lemma 5.1 *The restriction-induction chain on $\widehat{G}$ is reversible with respect to the Plancherel measure, that is,*

$$M_{\mathrm{Pl}}^G(\{\lambda\}) P_{\lambda\mu} = M_{\mathrm{Pl}}^G(\{\mu\}) P_{\mu\lambda}, \qquad \lambda, \mu \in \widehat{G}.$$

Hence the chain keeps M_{Pl}^G invariant.

Proof We immediately have

$$M_{\mathrm{Pl}}^G(\{\lambda\}) P_{\lambda\mu} = \frac{\dim \lambda \dim \mu}{|G|[G:H]} \sum_{\xi \in \widehat{H}} c_{\lambda,\xi} c_{\mu,\xi},$$

which is symmetric in λ and μ.

Let χ^Π denote the character of a representation Π of G on a finite-dimensional vector space V, $\tilde{\chi}^\Pi$ be the normalized one $\chi^\Pi/\dim V$, and χ_C^Π denote the value at an element of a conjugacy class C of G. If π is a representation of H, the induced character formula for $\Pi = \mathrm{Ind}_H^G \pi$ is well-known:

$$\tilde{\chi}^\Pi(x) = \frac{1}{|G|} \sum_{y \in G} \tilde{\chi}^\pi(y^{-1}xy), \qquad x \in G, \tag{5.3}$$

where χ^π is extended onto G by setting $\chi^\pi(x) = 0$ for $x \notin H$. For a conjugacy class C of G, decompose its restriction to H as $C \cap H = \bigsqcup_i C_i$ into conjugacy classes C_i of H. Then, (5.3) is rewritten as

$$\tilde{\chi}_C^\Pi = \sum_i \frac{|C_i|}{|C|} \tilde{\chi}_{C_i}^\pi. \tag{5.4}$$

We seek eigenvectors of the transition matrix $\mathbf{P} = [P_{\lambda\mu}]_{\lambda,\mu \in \widehat{G}}$ of the restriction-induction chain. Let $\tilde{\mathbf{x}}_C$ denote the column vector $[\tilde{\chi}_C^\lambda]_{\lambda \in \widehat{G}}$ for a conjugacy class C of G.

Lemma 5.2 *We have*

$$\mathbf{P}\,\tilde{\mathbf{x}}_C = \frac{|C \cap H|}{|C|}\,\tilde{\mathbf{x}}_C. \tag{5.5}$$

Proof The λ-entry of $\mathbf{P}\,\tilde{\mathbf{x}}_C$ for $\lambda \in \widehat{G}$ is computed as

$$\sum_{\mu \in \widehat{G}} P_{\lambda\mu}\tilde{\chi}_C^{\mu} = \frac{1}{[G:H]\dim\lambda} \sum_{\mu \in \widehat{G}}\sum_{\xi \in \widehat{H}} c_{\lambda,\xi}\, c_{\mu,\xi}\, \chi_C^{\mu}$$

$$= \frac{1}{[G:H]\dim\lambda} \sum_{\xi \in \widehat{H}} c_{\lambda,\xi}\, \chi_C^{\mathrm{Ind}_H^G \xi} = \frac{1}{[G:H]\dim\lambda}\, \chi_C^{\mathrm{Ind}_H^G \mathrm{Res}_H^G \lambda}. \tag{5.6}$$

Applying (5.4) with the decomposition $C \cap H = \bigsqcup_i C_i$, we have

$$\chi_C^{\mathrm{Ind}_H^G \mathrm{Res}_H^G \lambda} = \sum_i [G:H]\frac{|C_i|}{|C|}\chi_{C_i}^{\mathrm{Res}_H^G \lambda} = [G:H]\frac{|C \cap H|}{|C|}\chi_C^{\lambda}.$$

Hence (5.6) equals $(|C \cap H|/|C|)\tilde{\chi}_C^{\lambda}$. This completes the proof of (5.5). $\qquad\blacksquare$

Remark 5.1 Restriction-induction chains are effectively used in Fulman's works [9, 10] etc., which should have been mentioned in [12] also.

5.2 Diffusive Limit

Our Markov chain on $\mathbb{Y}_n$ mentioned in the beginning of Sect. 5.1 is produced by the restriction-induction chain for $G = \mathfrak{S}_n$ and $H = \mathfrak{S}_{n-1}$. In this situation, let us see the transition probability $P_{\lambda\mu}^{(n)}$ in (5.2) for $\lambda, \mu \in \mathbb{Y}_n$. Here the superscript $^{(n)}$ is put to make dependence on n explicit. We now have

$$c_{\lambda,\xi} = \begin{cases} 1, & \xi \nearrow \lambda, \\ 0, & \text{otherwise}, \end{cases} \qquad \lambda \in \mathbb{Y}_n, \ \xi \in \mathbb{Y}_{n-1}.$$

If $\lambda = \mu \in \mathbb{Y}_n$, (5.2) yields

$$P_{\lambda\lambda}^{(n)} = \frac{1}{[\mathfrak{S}_n : \mathfrak{S}_{n-1}]} \sum_{\xi \in \mathbb{Y}_{n-1}: \xi \nearrow \lambda} 1 = \frac{1}{n}\big|\{\text{peaks of the profile of }\lambda\}\big|.$$

If $\lambda, \mu \in \mathbb{Y}_n$ are distinct, there possibly exists at most one $\xi \in \mathbb{Y}_{n-1}$ such that $\xi \nearrow \lambda$ and $\xi \nearrow \mu$. This ξ is the set-theoretical intersection of the boxes of λ and μ: $\xi = \lambda \wedge \mu$. We thus gets

$$P_{\lambda\mu}^{(n)} = \begin{cases} |\{\text{peaks of the profile of } \lambda\}|/n, & \lambda = \mu, \\ \dim \mu/(n \dim \lambda), & \lambda \wedge \mu \in \mathbb{Y}_{n-1}, \\ 0, & \text{otherwise.} \end{cases} \tag{5.7}$$

Let us consider a continuous time Markov chain $(X_s^{(n)})_{s \in [0,\infty)}$ with the transition matrix $\mathbf{P}^{(n)} = [P_{\lambda\mu}^{(n)}]_{\lambda,\mu \in \mathbb{Y}_n}$ on the state space $\mathbb{Y}_n$. Let $\mathbb{M}_0^{(n)}$ be the initial distribution on $\mathbb{Y}_n$. The induced probability on the set of paths (namely, $\mathbb{Y}_n$-valued functions on $[0, \infty)$) is denoted by $\mathscr{M}^{(n)}$. Then, the distribution $\mathscr{M}^{(n)}(X_s^{(n)} = \cdot)$ at time s is given by

$$\mathscr{M}^{(n)}(X_s^{(n)} = \mu) = \sum_{\lambda \in \mathbb{Y}_n} \mathbb{M}_0^{(n)}(\{\lambda\})\left(e^{s(\mathbf{P}^{(n)}-\mathbf{I})}\right)_{\lambda\mu}, \qquad \mu \in \mathbb{Y}_n. \tag{5.8}$$

We take the limit of both s and n tending to ∞ in a *diffusive* regime, namely under the rescales of time and space in micro-macro transition:

$$s \in [0, \infty) \longmapsto \frac{s}{n}, \qquad \lambda \in \mathbb{Y}_n \longmapsto \lambda^{\sqrt{n}}(x) = \frac{1}{\sqrt{n}}\lambda(\sqrt{n}x) \tag{5.9}$$

respectively. Thus, for (macroscopic) time $t \in [0, \infty)$, let $\mathbb{M}_t^{(n)}$ be the distribution of the chain at (microscopic) time $s = tn$:

$$\mathbb{M}_t^{(n)}(\{\lambda\}) = \mathscr{M}^{(n)}(X_{tn}^{(n)} = \lambda), \qquad \lambda \in \mathbb{Y}_n. \tag{5.10}$$

The following result tells us that the concentration property of an initial state is propagated as macroscopic time goes by in our model. The examples mentioned in Sect. 4.4 serve to produce such initial states.

Theorem 5.1 *For a sequence of the Markov chains $\left\{(X_s^{(n)})_{s \in [0,\infty)}\right\}_{n \in \mathbb{N}}$, assume that the initial probability space $\left\{(\mathbb{Y}_n, \mathbb{M}_0^{(n)})\right\}_{n \in \mathbb{N}}$ admits the concentration at $\omega_0 \in \mathbb{D}$. Then, for any macroscopic time $t \in (0, \infty)$, $\left\{(\mathbb{Y}_n, \mathbb{M}_t^{(n)})\right\}_{n \in \mathbb{N}}$ also admits the concentration at some $\omega_t \in \mathbb{D}$. The limit profile ω_t is characterized through its transition measure $\mathfrak{m}_{\omega_t}$ by using the free convolution and the free compression:*

$$\mathfrak{m}_{\omega_t} = (\mathfrak{m}_{\omega_0})_{e^{-t}} \boxplus (\mathfrak{m}_\Omega)_{1-e^{-t}} \tag{5.11}$$

where Ω is the limit shape of (4.6) with the standard semi-circle distribution as its transition measure $\mathfrak{m}_\Omega$.

Proof [*Step* 1] We first translate (5.11) in terms of the free cumulant sequence. By (1.24) and (1.33), we see (5.11) is equivalent to

$$\begin{aligned} R_1(\mathfrak{m}_{\omega_t}) = 0, \qquad R_2(\mathfrak{m}_{\omega_t}) &= 1, \\ R_k(\mathfrak{m}_{\omega_t}) = R_k(\mathfrak{m}_{\omega_0})e^{-(k-1)t}, \qquad k &\in \{3, 4, \cdots\}. \end{aligned} \tag{5.12}$$

Note that $R_1(\mathrm{m}_{\omega_0}) = 0$, $R_2(\mathrm{m}_{\omega_0}) = 1$ hold by the assumption of concentration, especially (4.47).

[*Step* 2] We use Lemma 5.2 for $G = \mathfrak{S}_n$ and $H = \mathfrak{S}_{n-1}$. Let C be the conjugacy class of $\mathfrak{S}_n$ associated with $(\rho, 1^{n-|\rho|})$ for $\rho \in \mathbb{Y}$. Since

$$\frac{|C \cap \mathfrak{S}_{n-1}|}{|C|} = \frac{(n - |\rho| + m_1(\rho))!(n-1)!}{n!(n - 1 - |\rho| + m_1(\rho))!} = 1 - \frac{|\rho| - m_1(\rho)}{n}$$

holds, applying (5.5) to $\tilde{\mathbf{x}}_{(\rho,1^{n-|\rho|})} = \left[\tilde{\chi}^{\lambda}_{(\rho,1^{n-|\rho|})}\right]_{\lambda \in \mathbb{Y}_n}$, we have

$$\mathbf{P}^{(n)}\tilde{\mathbf{x}}_{(\rho,1^{n-|\rho|})} = \left(1 - \frac{|\rho| - m_1(\rho)}{n}\right)\tilde{\mathbf{x}}_{(\rho,1^{n-|\rho|})} \tag{5.13}$$

and hence

$$e^{tn(\mathbf{P}^{(n)}-\mathbf{I})}\tilde{\mathbf{x}}_{(\rho,1^{n-|\rho|})} = e^{-t(|\rho|-m_1(\rho))}\tilde{\mathbf{x}}_{(\rho,1^{n-|\rho|})}. \tag{5.14}$$

Among several criteria for the concentration developed in Sect. 4.4, we use the one described in terms of Σ_ρ's by taking (5.14) into account. Combining (5.10), (5.8) and (5.14), we obtain for $\rho \in \mathbb{Y}$

$$\begin{aligned}
E_{\mathbb{M}_t^{(n)}}[\Sigma_\rho] &= \sum_{\mu \in \mathbb{Y}_n} \mathbb{M}_t^{(n)}(\{\mu\}) \Sigma_\rho(\mu) \\
&= \sum_{\mu \in \mathbb{Y}_n} \left(\sum_{\lambda \in \mathbb{Y}_n} \mathbb{M}_0^{(n)}(\{\lambda\})\left(e^{tn(\mathbf{P}^{(n)}-\mathbf{I})}\right)_{\lambda\mu}\right) \Sigma_\rho(\mu) \\
&= \sum_{\lambda \in \mathbb{Y}_n} \mathbb{M}_0^{(n)}(\{\lambda\}) \sum_{\mu \in \mathbb{Y}_n}\left(e^{tn(\mathbf{P}^{(n)}-\mathbf{I})}\right)_{\lambda\mu} \Sigma_\rho(\mu) \\
&= \sum_{\lambda \in \mathbb{Y}_n} \mathbb{M}_0^{(n)}(\{\lambda\}) e^{-t(|\rho|-m_1(\rho))} \Sigma_\rho(\lambda) \\
&= e^{-t(|\rho|-m_1(\rho))} E_{\mathbb{M}_0^{(n)}}[\Sigma_\rho].
\end{aligned}$$

This implies that (4.51) satisfied by $\mathbb{M}_0^{(n)}$ is inherited by $\mathbb{M}_t^{(n)}$ also. Moreover,

$$n^{-(j+1)/2} E_{\mathbb{M}_t^{(n)}}[\Sigma_j] = e^{-jt} n^{-(j+1)/2} E_{\mathbb{M}_0^{(n)}}[\Sigma_j]$$
$$\xrightarrow{n \to \infty} e^{-jt} r_{j+1} = e^{-jt} R_{j+1}(\mathrm{m}_{\omega_0})$$

holds for $j \in \{2, 3, \cdots, \}$, which agrees with the free cumulant sequence of (5.12). We have thus shown $\left\{(\mathbb{Y}_n, \mathbb{M}_t^{(n)})\right\}_{n \in \mathbb{N}}$ admits the concentration at ω_t determined by (5.11).

Remark 5.2 If we adopt the induction-restriction chain instead of the restriction-induction one as a microscopic dynamics, (5.7) is replaced by

$$P_{\lambda\mu}^{(n)} = \begin{cases} \big|\{\text{valleys of the profile of } \lambda\}\big|/(n+1), & \lambda = \mu, \\ \dim \mu/((n+1)\dim \lambda), & \lambda \vee \mu \in \mathbb{Y}_{n+1}, \\ 0, & \text{otherwise}, \end{cases}$$

where $\lambda \vee \mu$ is the set-theoretical union of the boxes of λ and μ. Again, this chain is reversible with respect to $M_{\mathrm{Pl}}^{(n)}$. We modify (5.13) and (5.14) as

$$\mathbf{P}^{(n)} \tilde{\mathbf{x}}_{(\rho, 1^{n-|\rho|})} = \left(1 - \frac{|\rho| - m_1(\rho)}{n+1}\right) \tilde{\mathbf{x}}_{(\rho, 1^{n-|\rho|})},$$

$$e^{tn(\mathbf{P}^{(n)} - \mathbf{I})} \tilde{\mathbf{x}}_{(\rho, 1^{n-|\rho|})} = e^{-t\frac{n}{n+1}(|\rho| - m_1(\rho))} \tilde{\mathbf{x}}_{(\rho, 1^{n-|\rho|})}$$

respectively for $\rho \in \mathbb{Y}$ and $n \in \mathbb{N}$. Hence Theorem 5.1 remains valid without any modification.

By (3.26) and (5.12), we have

$$\int_{\mathbb{R}} (\omega_t(x) - |x|)dx = 2, \qquad t \in [0, \infty).$$

The macroscopic profile ω_t is regarded as the interface of the region between $y = \omega_t(x)$ and $y = |x|$ which has constant area 2. We see another aspect of the time evolution of ω_t in terms of the Stieltjes transform of its transition measure. Set

$$G(t, z) = G_{\mathfrak{m}_{\omega_t}}(z) = \int_{\mathbb{R}} \frac{1}{z - x} \mathfrak{m}_{\omega_t}(dx). \tag{5.15}$$

The following is a (nonlinear) PDE aspect of our dynamical model.

Theorem 5.2 *The function $G(t, z)$ of (5.15) satisfies the partial differential equation:*

$$\frac{\partial G}{\partial t}(t, z) = G(t, z) + \frac{1}{G(t, z)} \frac{\partial G}{\partial z}(t, z) - G(t, z) \frac{\partial G}{\partial z}(t, z). \tag{5.16}$$

Proof Considering (5.12) in (1.26) for $\mu = \mathfrak{m}_{\omega_0}$ and $\mu = \mathfrak{m}_{\omega_t}$, we have

$$K_0(\zeta) = K_{\mathfrak{m}_{\omega_0}}(\zeta) = \zeta^{-1} + \zeta + \sum_{k=2}^{\infty} R_{k+1}(\mathfrak{m}_{\omega_0})\zeta^k,$$

$$K(t, \zeta) = K_{\mathfrak{m}_{\omega_t}}(\zeta) = \zeta^{-1} + \zeta + \sum_{k=2}^{\infty} R_{k+1}(\mathfrak{m}_{\omega_0})e^{-kt}\zeta^k,$$

and hence

$$K_0(\zeta e^{-t}) = \zeta^{-1}e^t + \zeta e^{-t} + \sum_{k=2}^{\infty} R_{k+1}(\mathfrak{m}_{\omega_0})e^{-kt}\zeta^k$$

$$= \zeta^{-1}e^t + \zeta e^{-t} + K(t, \zeta) - \zeta^{-1} - \zeta. \tag{5.17}$$

Differentiate (5.17) in t and ζ respectively and eliminate the terms containing K_0'. Then,

$$\frac{\partial K}{\partial t}(t, \zeta) + \zeta \frac{\partial K}{\partial \zeta}(t, \zeta) + \zeta^{-1} - \zeta = 0. \tag{5.18}$$

On the other hand, we have

$$K\big(t, G(t, z)\big) = K_{\mathfrak{m}_{\omega_t}}\big(G_{\mathfrak{m}_{\omega_t}}(z)\big) = z, \tag{5.19}$$

and hence

$$\frac{\partial K}{\partial t}\big(t, G(t, z)\big) + \frac{\partial K}{\partial \zeta}\big(t, G(t, z)\big)\frac{\partial G}{\partial t}(t, z) = 0,$$

$$\frac{\partial K}{\partial \zeta}\big(t, G(t, z)\big)\frac{\partial G}{\partial z}(t, z) = 1 \tag{5.20}$$

by differentiating (5.19) in t and z. Replacing $\frac{\partial K}{\partial t}$ and $\frac{\partial K}{\partial \zeta}$ in (5.18) by the expressions obtained from (5.20), we have the desired Eq. (5.16).

Remark 5.3 As seen from (5.12), we have the convergence of moments and hence

$$\lim_{t \to \infty} \mathfrak{m}_{\omega_t} = \mathfrak{m}_{\Omega} \quad \text{in} \quad \mathscr{P}(\mathbb{R}).$$

On the other hand, the ODE

$$G(z) + \frac{1}{G(z)}\frac{dG(z)}{dz} - G(z)\frac{dG(z)}{dz} = 0$$

connected with (5.16) is easily solved to have the solution

$$G(z) = \frac{z - \sqrt{z^2 - 4}}{2},$$

which is the Stieltjes transform $G_{\mathfrak{m}_{\Omega}}(z)$ of $\mathfrak{m}_{\Omega}$ (see (3.16)).

Remark 5.4 Funaki–Sasada [11] gave remarkable results on hydrodynamic limit for the evolution of profiles of Young diagrams. Their model is given in the setting of the grand canonical ensemble. The Markov chain governing the microscopic dynamics runs over $\mathbb{Y}$, totality of Young diagrams of all sizes, allowing variation of the number of boxes. In one step transition from $\lambda \in \mathbb{Y}$, all peaks of λ are treated equally for removal of a box, and similarly all valleys for addition.

Remark 5.5 Borodin–Olshanski [5] showed a very interesting scaling limit for Markov chains on $\mathbb{Y}_n$ in a diffusive regime for time vs space. Their limit of $n \to \infty$ is taken not under the rescale in which the profile of a Young diagram survives but under the one in which characters of $\mathfrak{S}_\infty$ are captured, that is, under the famous Vershik–Kerov condition. Instead of (5.9), the rescales of time and space in micro-macro transition are given by $1/n^2$ and $1/n$ respectively for the size n of a Young diagram. The Markov chain governing the microscopic dynamics keeps a z-measure invariant. The constructed diffusion process on the Thoma simplex Δ has rich structure to be investigated.

References

1. Biane, P.: Representations of symmetric groups and free probability. Adv. Math. **138**, 126–181 (1998)
2. Biane, P.: Approximate factorization and concentration for characters of symmetric groups. IMRN **2001**, 179–192 (2001)
3. Biane, P.: Characters of symmetric groups and free cumulants. In: Vershik, A.M. (ed.) Asymptotic Combinatorics with Applications to Mathematical Physics, vol. 1815, pp. 185–200. Springer, Lecture Notes in Mathematics (2003)
4. Borodin, A., Bufetov, A., Olshanski, G.: Limit shapes for growing extreme characters of $U(\infty)$. Ann. Appl. Probab. **25**, 2339–2381 (2015)
5. Borodin, A., Olshanski, G.: Infinite-dimensional diffusions as limits of random walks on partitions. Probab. Theory Relat. Fields **144**, 281–318 (2009)
6. Ceccherini-Silberstein, T., Scarabotti, F., Tolli, F.: Representation Theory of the Symmetric Groups. Cambridge University Press, Cambridge studies in advanced mathematics (2010)
7. Chandrasekharan, K.: Arithmetical Functions. Springer (1970)
8. Féray, V.: Combinatorial interpretation and positivity of Kerov's character polynomials. J. Algebr. Comb. **29**, 473–507 (2009)
9. Fulman, J.: Stein's method, Jack measure, and the Metropolis algorithm. J. Combin. Theory Ser. A **108**, 275–296 (2004)
10. Fulman, J.: Stein's method and Plancherel measure of the symmetric group. Trans. Amer. Math. Soc. **357**, 555–570 (2005)
11. Funaki, T., Sasada, M.: Hydrodynamic limit for an evolutional model of two-dimensional Young diagrams. Commun. Math. Phys. **299**, 335–363 (2010)
12. Hora, A.: A diffusive limit for the profiles of random Young diagrams by way of free probability. Publ. RIMS Kyoto Univ. **51**, 691–708 (2015)
13. Hora, A.: Representations of Symmetric Groups and Analysis of Ensembles of Young Diagrams (in Japanese). Sugaku Shobo, To appear
14. Hora, A., Obata, N.: Quantum Probability and Spectral Analysis of Graphs. Springer, Theoretical and Mathematical Physics (2007)
15. Ivanov, V., Kerov, S.: The algebra of conjugacy classes in symmetric groups and partial permutations. J. Math. Sci. **107**, 4212–4230 (2001)
16. Ivanov, V., Olshanski, G.: Kerov's central limit theorem for the Plancherel measure on Young diagrams. In: Fomin, S. (ed.) Symmetric Functions 2001: Surveys of Developments and Perspectives, vol. 74, pp. 93–151. Physics and Chemistry. Kluwer Academic Publishers, NATO Science Series II, Mathematics (2002)
17. Kerov, S.: Gaussian limit for the Plancherel measure of the symmetric group. C.R. Acad. Sci. Paris Sér. I Math. **316**, 303–308 (1993)
18. Kerov, S.: Interlacing measures. Amer. Math. Soc. Transl. **181**, 35–83 (1998)

© The Author(s) 2016
A. Hora, *The Limit Shape Problem for Ensembles of Young Diagrams*,
SpringerBriefs in Mathematical Physics, DOI 10.1007/978-4-431-56487-4

19. Kerov, S.V.: Asymptotic Representation Theory of the Symmetric Group and its Applications in Analysis, Translations of Mathematical Monographs, vol. 219. American Mathematical Society (2003)
20. Kerov, S., Olshanski, G.: Polynomial functions on the set of Young diagrams. C.R. Acad. Sci. Paris Sér. I Math. **319**, 121–126 (1994)
21. Logan, B.F., Shepp, L.A.: A variational problem for random Young tableaux. Adv. Math. **26**, 206–222 (1977)
22. Macdonald, I.G.: Symmetric Functions and Hall Polynomials, 2nd Ed. Oxford University Press (1995)
23. Nica, A., Speicher, R.: Lectures on the Combinatorics of Free Probability, London Mathematical Society and Lecture Note Series, vol. 335. Cambridge University Press (2006)
24. Okada, S.: Representation Theory of Classical Groups and Combinatorics, vols. 1, 2 (in Japanese). Baifukan (2006)
25. Sagan, B.E.: The Symmetric Group: Representations, Combinatorial Algorithms, and Symmetric Functions, 2nd Ed., Graduate Texts in Mathematics, vol. 203. Springer (2001)
26. Śniady, P.: Gaussian fluctuations of characters of symmetric groups and of Young diagrams. Probab. Theory Relat. Fields **136**, 263–297 (2006)
27. Terada, I., Harada, K.: Group Theory (in Japanese). Iwanami Shoten (1997)
28. Thoma, E.: Die unzerlegbaren positiv-definiten Klassenfunktionen der abzählbar unendlichen, symmetrischen Gruppe. Math. Z. **85**, 40–61 (1964)
29. Vershik, A.M., Kerov, S.V.: Asymptotics of the Plancherel measure of the symmetric group and the limiting form of Young tables. Soviet Math. Dokl. **18**, 527–531 (1977)
30. Vershik, A.M., Kerov, S.V.: Asymptotic theory of characters of the symmetric group. Funct. Anal. Appl. **15**, 246–255 (1981)
31. Vershik, A.M., Kerov, S.V.: Asymptotic of the largest and the typical dimensions of irreducible representations of a symmetric group. Funct. Anal. Appl. **19**, 21–31 (1985)
32. Voiculescu, D.V., Dykema, K.J., Nica, A.: Free random variables, CRM Monograph Series. Amer. Math. Soc. **1** (1992)

Index

© The Author(s) 2016
A. Hora, *The Limit Shape Problem for Ensembles of Young Diagrams*,
SpringerBriefs in Mathematical Physics, DOI 10.1007/978-4-431-56487-4

Printed in the United States
By Bookmasters